Carcinogens and Mutagens in the Environment

Volume IV
The Workplace: Monitoring and Prevention of Occupational Hazards

Editor

Hans F. Stich, Ph.D.
Head
Environmental Carcinogenesis Unit
British Columbia Cancer Research Centre
Professor of Zoology
University of British Columbia
Vancouver, British Columbia, Canada

CRC Press, Inc.
Boca Raton, Florida

Library of Congress Cataloging in Publication Data

(Revised for volumes 4 and 5)
Main entry under title:

Carcinogens and mutagens in the environment.

Includes bibliographies and indexes.
Contents: v. 1. Food products -- [etc.] -- v. 4. The workplace: monitoring and prevention of occupational hazards -- v. 5. The workplace: sources of carcinogens.
1. Carcinogens. 2. Mutagens. 3. Environmentally induced diseases. I. Stich, H. F. (Hans F.). 1927-
[DNLM: 1. Carcinogens, Environmental. 2. Mutagens. 3. Environmental pollutants. QZ 202 C2651]
RC268.6.C365 1982 616.99'4071 81-21764
ISBN 0-8493-5881-7 (v. 1)

Direct all inquiries to CRC Press, Inc., 2000 Corporate Blvd., N.W., Boca Raton, Florida, 33431.

International Standard Book Number 0-8493-5884-1 (Volume IV)

Library of Congress Card Number 81-21764
Printed in the United States

PREFACE

Daily each member of the human population ingests, inhales, or comes into contact in other ways with thousands of chemicals, including carcinogens, mutagens, and teratogens. The individual is constantly bombarded by news that one or more of the most cherished food items is hazardous to health. Regulatory agencies face the difficulty each day of making decisions on the basis of totally inadequate scientific data. It is therefore not surprising to find a frightened public, irritated industrialists, pressed civil servants, and underpaid scientists clamoring for more data on human populations and larger influence in the decision-making process that may affect everyone's lives.

Our lack of understanding of environmental toxicology becomes only too evident from the observation that we cannot even guess about the relative contribution of man-made agents and naturally occurring chemicals to the total carcinogenic or mutagenic load in a particular environment. The information available to assess a particular hazard to human health is shockingly inadequate when we consider the millions of chemicals in our environment, the few thousand chemicals tested for genotoxicity, and the paltry hundred or so compounds that have been properly evaluated for their carcinogenicity and teratogenicity. The difficulties involved in trying to assess the carcinogenic and genotoxic load will be further compounded when testing programs are extended to complex mixtures. The carcinogenicity and genotoxicity of a complex mixture does not seem to be a simple sum of the activities of the individual compounds because of a multitude of synergistic, antagonistic, co-carcinogenic, and co-mutagenic interactions between its various components.

Moreover, the composition of any mixture when ingested or inhaled will be subject to continuous change within the body that will in turn affect its carcinogenic and genotoxic potential. Finally, the testing of combinations and permutations between only a few compounds would rapidly exhaust the total available manpower and funding of an entire nation. A simple calculation reveals that the testing of all permutations between 10 different chemicals would lead to about 10 million possible permutations, $1^1/_2$ billion dishes, approximately 79 million working hours, and 9.8×10^9 U.S. dollars when the Ames' Salmonella mutagenicity test is used on 5 strains, at 5 doses, and with or without S9 activation. With 20 different compounds, the number of man years would amount to approximately 3.3×10^{16} and the cost to 6.5×10^{21} U.S. dollars. These are staggering figures considering that 20 compounds is a relatively small number when we realize that 2000 or more chemicals are found, for example, in such widely consumed beverages as coffee and tea. In spite of these and other apparently insurmountable difficulties, the search for solutions will and must continue. The papers in this volume are an example of this effort.

H.F. Stich

THE EDITOR

Hans F. Stich, Ph.D., is Head, Environmental Carcinogenesis Unit, British Columbia Cancer Research Centre and Professor in the Department of Zoology and in the Department of Pathology, University of British Columbia, Vancouver, B. C., Canada.

Dr. Stich received his Ph.D. in Developmental Biology in 1949 from Faculty of Arts and Science at the University of Wurzburg, Germany. He also studied tropical medicine at the same university.

He is a member of the Environmental Contaminants Advisory Committee on Mutagenesis, Health and Welfare Canada; a Director of the British Columbia Cancer Foundation; Professional Consultant of the B.C. Cancer Control Agency; Councilor of the Environment Mutagen Society; on the editorial board of *Mutation Research* and *Cancer Research*; was chairman of the Grant Panel on Environment Toxicology; and Chairman of the Grant Panel on Cell Biology and Genetics of the Natural Science and Engineering Research Council Canada.

Dr. Stich has presented numerous invited lectures at international and national meetings as well as guest lectures at various universities and institutes. He has published over 180 papers to date. His major research interests include the etiology of oral and esophageal carcinomas among betel nut and tobacco chewers, the use of genotoxicity tests to detect naturally occurring carcinogens in food products and beverages, and the assessment of a mutagenic load by examining the synergistic and antagonistic interactions between mutagens, co-mutagens, and antimutagens.

CONTRIBUTORS

Lorenzo Alessio, M.D.
Associate Professor
Institute of Occupational Medicine
Clinica del Lavoro "L. Devoto"
University of Milan
Milan, Italy

June J. Andersen, Ph.D.
Manager
Advanced Disk Product Engineering
General Products Division
IBM Corporation
San Jose, California

Gordon R. C. Atherley, M.D., LL.D.
President and Chief Executive Officer
Canadian Centre for Occupational Health and Safety
Hamilton, Ontario, Canada

R. A. Baan, Ph.D.
Biochemist
Department of Genetic Toxicology
Medical Biological Laboratory TNO
Rijswijk, The Netherlands

Ann D. Burrell
General Products Division
IBM Corporation
San Jose, California

Gordon C. Butler
Division of Biological Sciences
National Research Council of Canada
Ottawa, Ontario, Canada

Andrew Churg, M.D.
Associate Professor
Department of Pathology
University of British Columbia
Vancouver, British Columbia, Canada

D. E. W. Clarke
Rank-Xerox Limited
Welwyn Garden City, England

Betty J. Dabney
General Products Division
IBM Corporation
Boulder, Colorado

Gary M. Decad
General Products Division
IBM Corporation
San Jose, California

C. J. DeMarco
Corporate Environmental Health and Safety Department
Xerox Corporation
Rochester, New York

A. M. J. Fichtinger-Schepman, Ph.D.
Biochemist
Department of Genetic Toxicology
Medical Biological Laboratory TNO
Rijswijk, The Netherlands

Alessandra Forni, M.D.
Assistant Professor
Institute of Occupational Medicine
Clinica del Lavoro "L. Devoto"
Professor of Histology and General Embryology
University of Milan
Milan, Italy

Peter F. Infante, D.D.S., Dr.P.H.
Director
Office of Carcinogen Identification and Classification
Directorate of Health Standards Programs
Occupational Safety and Health Administration
Washington, D.C.

Göran Löfroth
Professor of Environmental Health
Nordic School of Public Health
Gothenburg, Sweden

P. H. M. Lohman, Ph.D.
Head
Department of Genetic Toxicology
Medical Biological Laboratory TNO
Rijswijk, The Netherlands

J. C. MacKenzie
Corporate Environmental Health and Safety Department
Xerox Corporation
Rochester, New York

John C. Marshall, M.D.
Victoria General Hospital
Halifax, Nova Scotia, Canada

Robert Mermelstein, Ph.D.
Principal Scientist
Manager
Materials Environmental Health and Safety
Xerox Corporation
Rochester, New York

Anthony B. Miller, M.B., F.R.C.P.(C)
NCIC Epidemiology Unit
Faculty of Medicine
University of Toronto
Toronto, Ontario, Canada

T. J. Roberts
Corporate Environmental Health and Safety Department
Xerox Corporation
Rochester, New York

Ann E. Robinson, Ph.D., F.C.I.C.
Assistant Deputy Minister
Occupational Health and Safety Division
Ontario Ministry of Labour
Toronto, Ontario, Canada

Herbert S. Rosenkranz, Ph.D.
Professor
Director
Center for the Environmental Health Services
Case Western Reserve University
Cleveland, Ohio

Miriam P. Rosin, Ph.D.
Environmental Carcinogenesis Unit
British Columbia Cancer Research Centre
Vancouver, British Columbia, Canada

Marvin Schneiderman, Ph.D.
Department of Preventive Medicine and Biostatistics
Uniformed Services University of the Health Sciences
Bethesda, Maryland

M. A. Schoen
Biologist
Department of Genetic Toxicology
Medical Biological Laboratory TNO
Rijswijk, The Netherlands

H. H. Schutte, Ph.D.
Biochemist
Department of Genetic Toxicology
Medical Biological Laboratory TNO
Rijswijk, The Netherlands

Emmanuel Somers, Ph.D.
Director General
Environmental Health Directorate
Health and Welfare Canada
Health Protection Branch
Ottawa, Ontario, Canada

G. P. Van der Schans, Ph.D.
Biochemist
Department of Genetic Toxicology
Medical Biological Laboratory TNO
Rijswijk, The Netherlands

R. A. Wadden, Ph.D., P.E.
Professor and Director
Environmental and Occupational Health Studies
School of Public Health
University of Illinois
Chicago, Illinois

Jonathan B. Ward, Jr., Ph.D.
Assistant Professor
Department of Preventive Medicine and Community Health
Division of Environmental Toxicology
University of Texas Medical Branch
Galveston, Texas

Robert Whiting, Ph.D.
Industrial Toxicologist
Technical Services Section
Canadian Centre for Occupational Health
and Safety
Hamilton, Ontario, Canada

Gary M. Williams, M.D.
Associate Director
Naylor Dana Institute
American Health Foundation
Valhalla, New York

O. Brocades Zaalberg, M.D., Ph.D.
Head
Department of Immunology
Medical Biological Laboratory TNO
Rijswijk, The Netherlands

TABLE OF CONTENTS

Part A
Cancer As An Occupational Hazard

Chapter 1

OCCUPATIONAL HAZARDS: AN OVERVIEW

Marvin A. Schneiderman

TABLE OF CONTENTS

"Why did God create Adam alone?
In order to teach us that whoever destroys a single life is as guilty as though he had destroyed the entire world; and that whoever saves one life, earns as much merit as though he has saved the entire world."

The Talmud: Sanhedrin 4:5

I. INTRODUCTION

In the early spring of 1981, a Banbury conference was held on the subject "Quantification of Occupational Cancer". The proceedings have since been published.[1] A recent reviewer,[2] in addition to remarking on the unlikelihood that all increases were cigarette-related, also remarked, "quantification of occupational cancer is often less important than defining the risk of certain occupations and appropriately intervening once the risk has been established", with which the author agrees.

This paper concerns itself first with how the risk may be established. It describes a possible sequence of extrapolation which involves testing, observation, and some difficult data gathering (or nongathering) that should lead to establishing risk. In doing this the author follows some recent work of the NAS/NRC of the U.S.[3] This paper describes several of the rivers that must be crossed, and a few ingenious attempts to build useable bridges, if we are to cross to where we learn enough to take appropriate action.

Several estimates of the proportionate burden of industrially related cancers are compared and the impact is computed of a 5% total risk on the proportion of the population actually exposed.

Some consideration is given to the intent of some of the laws and the relation of this intent to survival of men in various occupations, and the association between social class and cancer mortality.

Some trends in cancer mortality among men in the U.S. are examined, raising more questions about what contributes to cancer mortality. These relationships are then looked upon as a possible basis for setting goals for the reduction of illness and death and the improvement of the working man's lot.

II. WHAT DOES ONE NEED TO KNOW?

The NAS/NRC[3] delineates four steps in the characterization of risk. The first is identification of the hazard; the second is quantification (i.e., dose-response) of the hazard; the third is exposure assessment, so that exposure (dose), times numbers exposed, and times response per unit dose can give a total estimated response, which then leads to the fourth step, "risk characterization." The NAS/NRC says this is "...the estimate of magnitude of the public health problem (and) involves no additional scientific knowledge or concepts." I suspect that unfortunately even after following this procedure, the risk characterization will read something like "...it may be many, it may not. The data are so sparse, so inadequately collected, confounded, confused, messy, missing or otherwise short of perfect that we'll have a very hard time telling."

Identification of hazards does not seem to be outrageously difficult. Using the data intelligently, (step 4) may be much harder. Laboratory studies have the potential for uncovering many carcinogens. From the short-term tests which seem to turn up many more positives than the large scale animal tests (in turn they may also turn up more "positive" materials than we are able to demonstrate in human studies), to the epidemiology studies which turn up positives (like benzene and arsenic) that the animal studies do not, there seems to be no lack of candidate materials to be considered hazards.

The problem of misclassification, however, is quite important, and no system has yet been devised which is without error. As an aside, it is perhaps important to consider misclassification for a moment. For example, there are 100,000 materials (chemicals, etc.) in commercial use (the true number may be nearer 65,000 or 70,000, but arithmetic with 100,000 is easier). Further, 10% of these, or 10,000 are really hazardous (carcinogens?) to humans. Finally, we have a superb set of laboratory tests — inexpensive, easily performed, which correctly identifies 90% of the positives as positives, and misclassifies as positives only 5% of the true negatives. This gives:

		Test results	
Real world		**Positive**	**Negative**
True +	10,000	9,000	1,000
True −	90,000	4,500	85,000
		13,500	86,500

Therefore, of 13,500 positive test results from this battery of tests (which is surely better than anything we now have) one third are false positive (4,500/13,500). Of the 86,500 test negatives, 1000, or 1.16% are false negatives. Depending on where one sits, a false negative may be much worse than a false positive, or a false positive much worse than a false negative. If you are the anti-aircraft crew and sit on the ground, false negatives in identifying enemy aircraft could kill you. If you are the flight crew and sit in the airplane while in flight, false positives (identifying you as enemy when you are not) could kill you. Now, equate the anti-aircraft crew to an exposed consumer or worker, and the flight crew to a manufacturer or seller. If you are a manufacturer or seller or developer of materials, false positives are anathema to you. If you are a user or are exposed to the materials, false negatives which get into commerce and may get into you, could kill you. If you are a government administrator or regulator and you have to solve the equation "how many false positives equal one false negative?", you should be prepared to realize that no answer will be universally accepted as correct.

III. EPIDEMIOLOGY IN HAZARD IDENTIFICATION

Because of this concern about misleading laboratory data (mostly small animal data) a polarization has developed. Some people want to depend more on short-term tests, and others, including some of the courts, want to base action largely on human data: epidemiology. As Karstadt points out however, epidemiology is unlikely to be done even for materials in extensive commercial use that are found to be carcinogenic in laboratory studies.[4] In view of this, new approaches to uncovering industrial hazards need to be developed that have more potential for discovering hazards with fewer false positives and false negatives.

Because cancer of a specific site is usually a rare disease, the probability of finding a link between a specific cancer and specific exposures is rather small. In addition, humans are rarely exposed to only one potential carcinogen, so that even after an epidemiological study is done, and excess risk is discovered, it can usually be said, "but those people were not only exposed to omega (the last word in carcinogens) but also alpha, beta, and gamma, so how can you be sure it was omega and not one or all of the others?" (A distinguishing characteristic of an epidemiologist is that he/she can always find something wrong with anything done by another epidemiologist.)

Following the lead of some of the British workers,[5] Gilbert Beebe[6] has suggested a system of record linkage, tying together different large record-keeping systems, such as the Census, a national death index, and death certificate reports that perhaps might improve the process.

It is interesting that several Canadians are doing some of the things Beebe proposed. Martha Smith,[7] from Statistics Canada, is one who has described the potential for long-term medical follow-up in Canada.

Some of what Smith discussed appears to have been put to use by Howe and Lindsay[8] in a first review of cancer mortality in men for the years 1965 to 1973, relating this mortality to occupational histories gathered in 1965 to 1969. The system has just begun to develop some momentum, and the authors are encouraged by their early results which confirm some previously suggested associations between occupation in some industries and cancer.

The Howe-Lindsay studies, based on a 10% sample of the Canadian labor force, are industry-oriented. That is, they look for elevated rates in an industry as a starting point to find what (if anything) in the industrial exposure might lead to such an increase. Siemiatycki and his group[9-11] from Montreal point out the problems (mostly dilution) in industry or area approaches (''Even if a high risk industry could be pinpointed with such an insensitive approach'' — here they were talking of geographically based studies — ''identification of the carcinogen(s) responsible would usually require additional research'') and have proposed a case-control-based system. They start with all cases of cancer and attempt to identify occupation for each case and possible carcinogen exposure (using the expertise of people called ''chemist-engineers'') by translating job histories into exposure history. In this manner, they believe they should be able to identify exposures that might occur in several jobs and in several industries. The preliminary work of this group has encouraged them to continue in their work. They are well aware of the problems involved in using as they do, other cancer cases as controls in part, expecting that this could lead to underestimation of risk. Although they mention a statistical technique by Duncan Thomas that should minimize this downward bias, some preliminary data they have published[11] on smoking and lung cancer seems to support this expectation of a downward bias. The slopes of the dose-response (relative risk: RR) curve they found for cigarette smoking and lung cancer was shallower than those reported by Doll and Hill,[12] Hammond and Horn,[13] and Dorn.[14] Crudely, for the U.S. data, $RR = 1 + d/2$, where d is the number of cigarettes smoked per day. Siemiatycki's data would lead to $RR = 1 + d/5$ or $1 + d/6$. This is a dose-response for smoking similar to one found in cigarette-smoking asbestos miners. Is lung cancer possibly being under-diagnosed in Quebec? Are Quebecois a ''resistant'' people?

Hoar[15] and colleagues at Harvard and the National Cancer Institute have used a system of classifying job titles and industry affiliations as a possible first step in the search for high-risk jobs (in contrast to Siemiatycki's chemist-engineer approach). Hoar reported,[15] ''we used the linkage system to convert the industry and occupational titles into a variable indicating exposure to aromatic amines. Our analysis (of bladder cancer cases) yielded higher risk estimates than the original analysis.''

IV. ESTIMATES OF INDUSTRIAL CANCER: SOURCES OF CONTROVERSY

Table 1 gives six different estimates of the proportion of cancers attributed to various factors, two of them including only occupation and exposure to industrial materials. The occupation estimates range from 1% in women to 25% of the cancers in men engaged in West German industry who were in contact with chemicals. The estimates were made in several different ways. Some authors tried to compute them directly from relative risks, persons exposed, and rates for nonexposed persons, as one would do if the following NAS/NRC 1983 approach. Some computations were done by looking at specific diseases and estimating the proportion of each disease with proved occupational associations with cancer. Some were done by subtraction, forcing a total to 100% after other ''causes'' were allocated. A few were done in keeping with Richard Doll's observation in 1967 (Rock Carling lecture ''Prevention of Cancer, Pointers from Epidemiology'') that probably 50% of lung cancers were associated with additional exposures to other carcinogens.

Table 1
PROPORTION OF CANCER DEATHS ATTRIBUTED TO VARIOUS FACTORS

	Doll and Peto[22]	Higginson and Muir[24]	Wynder[a] and Gori[25] ♂	Wynder[a] and Gori[25] ♀	Bridbord[a] et al.[16]	DFG[b]
Tobacco	30	30	28	8		
Alcohol	3	5	4	0.5		
Occupation	4	6	4	1.0	12.0	25[c]
Industrial products	1					
Medicines etc.	1	1 }	8	10.5		
Geophysical	3	11 }				
Infection	10					
Lifestyle factors		30				
Diet	35		41	60		
Reproductive & sexual behavior	7					
Food additives	1					
Pollution	2					
Others						
Congenital		2				
Unknown	?	15	15	20		

[a] Incidence.
[b] *Deutsche Forschungsgemeinschaft.*[17]
[c] Of men employed in contact with chemicals.

In looking at Table 1, it is worthwhile to inquire as to how the last two groups arrived at their estimates for industrial exposures, which are the highest among the five. Table 2 is a breakdown for the column headed "Bridbord et al." Like the Siemiatycki approach it is materials-oriented, under the assumption that what one can control is a material, not an industry, or a geographic area. It is reproduced from a document submitted from the testimony of Bridbord et al.[16] which was published (unedited) as an appendix to Banbury Report, No. 9.* Using the procedure later advocated by NAS/NRC, Bridbord and co-workers found for the six materials listed, that 46,900 new cases (deaths?) were anticipated annually. This amounts to about 12% of the annual deaths from cancer anticipated in the U.S. for 1980[1].

The German data are from a study by Horbach and Loskant[17] on workers in contact with chemicals in 10 German firms, involving 5104 cancer cases. Attempts were made to reconstruct the working life of over 40,000 employees. Horbach and Loskant concluded that about 25% of "all tumors which occurred in workers were due to influences at work." They added that "the overall risk to the population from products from the chemical industry is not as high as is widely thought." They also included the caveat that many of the deaths which occurred between 1950 and 1968 must have come from contacts with chemicals between the two world wars, when smaller quantities of chemicals were produced and safety precautions were fewer.

Table 3 shows how a low percent in the total population of cancer attributed to industrial exposure will lead to a high percent of cancers in the exposed populations. The data in Table 3 are crude, but are not grossly out of order, and show how an estimate of 5% of cancers in the total population translates to 25% of cancers in an exposed population.

* The body of the Bridbord testimony gives another, higher estimate based on a reconciliation of the asbestos mortality data with an earlier estimate by Selikoff, which anticipated 50,000 asbestos-related cancer deaths per year.

Table 2
NEW CANCER CASES RELATED TO INDUSTRIAL EXPOSURE IN THE U.S.[16]

Material	Persons exposed (NOHS)	Rel. risks (Cole-Goldman, Lassiter)	Diseases	Anticipated cases/yr
Asbestos	1,600,000	1.5—12	Lung Mesothelioma	13,900
Arsenic	1,500,000	3—8	Respiratory Skin (?)	7,300
Benzene	2,000,000	5	Leukemia Nasal	1,400
Chromium	1,500,000	3—40	Lung	7,900
Nickel	1,400,000	5—10	Respiratory Nasal (?)	7,300
Petroleum	3,900,000	2—33	Lung	9,100

Table 3
THE EFFECTS OF DILUTION

"Only 5% of cancer is related to occupation"

1. Because of employment practices, most of this will be in men.
   Blow-up factor: × 1.8 = 9.0
2. Most of this will come in the mining and manufacturing industries; about half, or less, of employment among men is in mining and manufacturing (e.g., 45%)
   Blow-up factor: × 2.2 = 19.8
3. Exposures are likely for about 3/4 of the work force in mining and manufacturing.
   Blow-up factor: × 1.3 = 25.7

Summary:
About 25% of cancers in exposed workers are related to their occupations (see results of DFG study).

Davis and Magee[18] have reported on the almost exponential growth in the production of "bio-available" chemicals since World War II. If persons are or were exposed to these materials on the job, Davis remarks, first that it will be hard to find unexposed populations, making it difficult to do epidemiologic studies, and further "...if the incidence pattern is similar to that between lung cancer and cigarette smoking, it may be expected that the 1980s, 1990s and the turn of the century will show the effect of increases in exposures that occurred in the 1960s."[19] Hoerger[20] has argued that although production has increased exponentially, exposure has not increased similarly, and because of better work practices in the large chemical companies in contrast to the exposures found in the converted service station chemical plants, exposures may have actually decreased.

There is some recent evidence showing changes in cancer mortality that may support Davis' perception. Overall, cancer mortality on an age-standardized basis increased rapidly in the first half of this century, then appeared to be flattening out in the 1950s and early 1960s, and since about 1960 to 1965 has once again been increasing. Some computations I[21] did recently indicated that had the data stopped in 1960, and had I fitted a second degree curve for the whole century up to 1960, I would have anticipated a maximum (age standardized to 1940) rate of 125.8 deaths per 100,000, reached at about 1962 or 1963. This rate has now been substantially exceeded (about 132/100,000 in 1981) and has not shown any signs of decline or leveling off in the last decade.

Total cancer mortality trends are composed of the trends in many diseases, and the trends

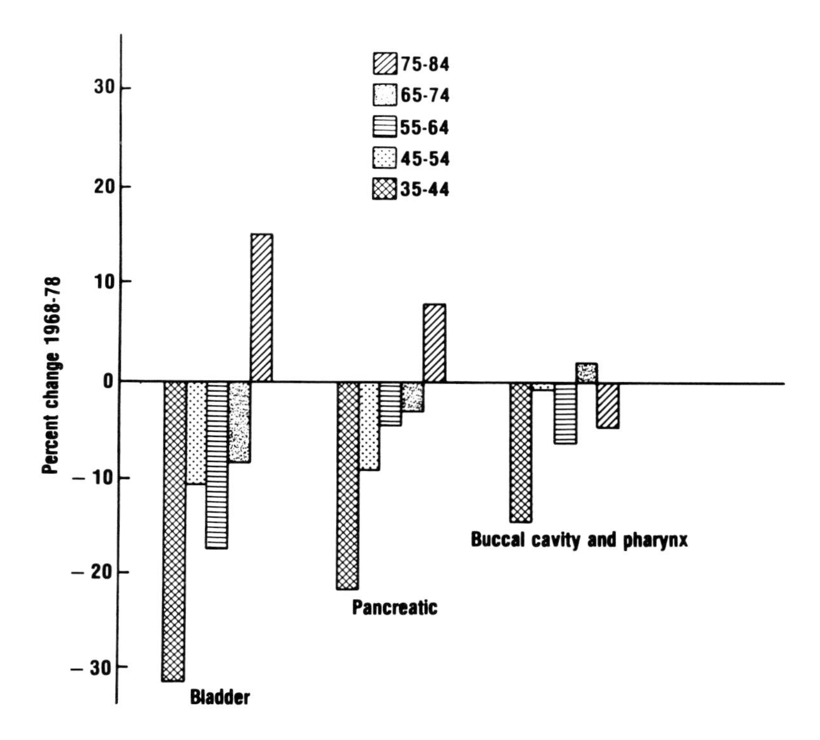

FIGURE 1. Cluster A: average percent change from 1968 to 1978 for cancer mortality for bladder, pancreas, and buccal cavity and pharynx for men aged 35 to 84 years.

of different age groups. Thus, mortality and incidence have been declining for a long time for stomach cancer and increasing for lung cancer. In fact, some people believe that the only increases of consequence in cancer incidence and mortality have come about in the cigarette-related cancers.[22] It therefore becomes worthwhile to examine some of the smoking-related cancers and some others in men where occupational exposure is most likely to be important. Davis et al.[23] have done this for nine primary sites in males age 35 to 84 in the U.S. from 1968 to 1978.

Two striking things may be seen in the data assembled by Davis. First, there are substantially different age-specific trends. For many of the cancers, the trends in mortality are down for the youngest persons (usually under age 55). Second, when the age-specific, site-specific changes from 1968 to 1978 are taken into account, the cancers appear to segregate themselves into three clusters: Cluster A — Cancers with declines in mortality for almost all except the oldest age group(s): bladder, pancreas, buccal cavity, and pharynx (Figure 1); Cluster B — Cancers with increases in almost all age groups: kidney, esophagus, liver (Figure 2); Cluster C — Cancers with decreases (or no increases) in younger age groups, but with very large percent changes (50% increase or more) in the oldest age group, 75 to 84: brain, multiple myeloma, and lung (Figure 3).

The most interesting aspect of this segregation of cancers into clusters with similar age-specific trends is that the smoking-related cancers — lung, bladder, buccal cavity and pharynx, pancreas, and possibly kidney — do not themselves constitute a cluster. Trends in lung cancer, the primary cigarette smoking-related cancer, behave more like the trends for brain cancer and multiple myeloma than bladder or buccal cavity or esophageal cancer. Because the percentage changes for age-specific groups are so dissimilar, it is hardly likely that the increases reported in brain cancer and multiple myeloma can reasonably all be

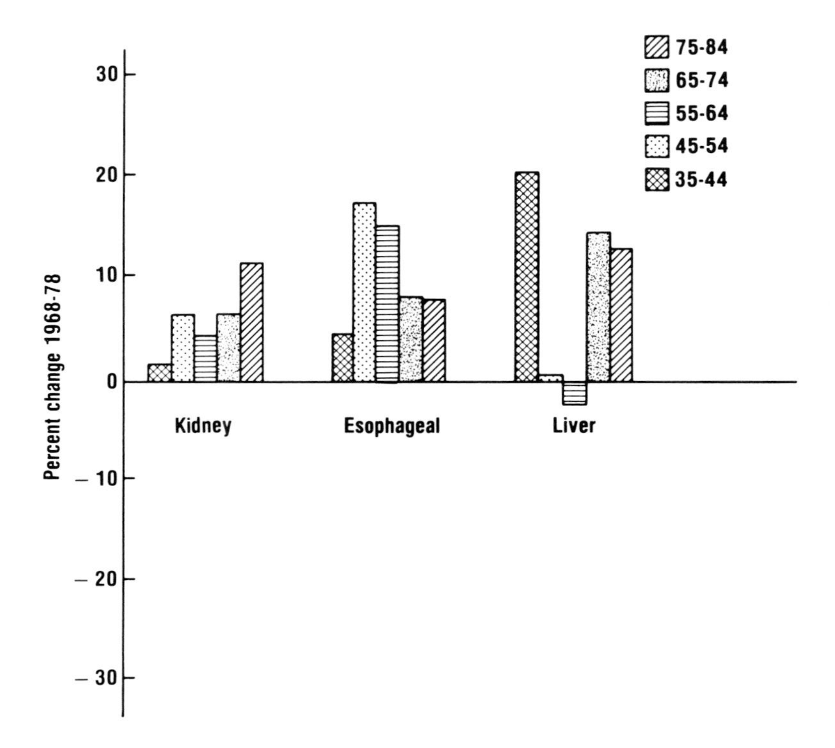

FIGURE 2. Cluster B: average percent change from 1968 to 1978 for cancer mortality for kidney, esophagus, and liver for men aged 35 to 84.

attributed to improvements in diagnosis. (Why would diagnosis improve more for old people than for young people?) In fact, such an argument would seem to lead to questioning whether all the increases in lung cancer in older men were also real, or whether some of the increases might be due to misdiagnosis or overdiagnosis. If one does not fully accept this improved diagnosis-misdiagnosis-overdiagnosis argument, then one may be more open to the suggestion that some of the increases seen may be related to industrial or general environmental exposure and not solely to cigarette smoking.

Having possibly convinced ourselves once again that industrial exposures are of importance to the life and health of industrial workers, it becomes important to ask what kind of meaningful goals should there be in attempting to improve the health of potentially exposed persons.

Following is some of the language in the U.S. occupational safety and health legislation, as an example of a possible goal. One section of the Occupational Safety and Health Act reads:

"To assure, so far as possible, every working man and woman in the nation safe and healthful working conditions and to preserve our human resources...
(7) by providing medical criteria which will assure
insofar as practicable that no employee will suffer
diminished health, functional capacity, or life
expectancy as a result of his work experience."

Figures 4 and 5 show differential mortality associated, first with social class (Figure 4) and then with job classification (Figure 5). Both are British data and are derived largely from a publication of the Government Statistical Service (Occupational Mortality: The Re-

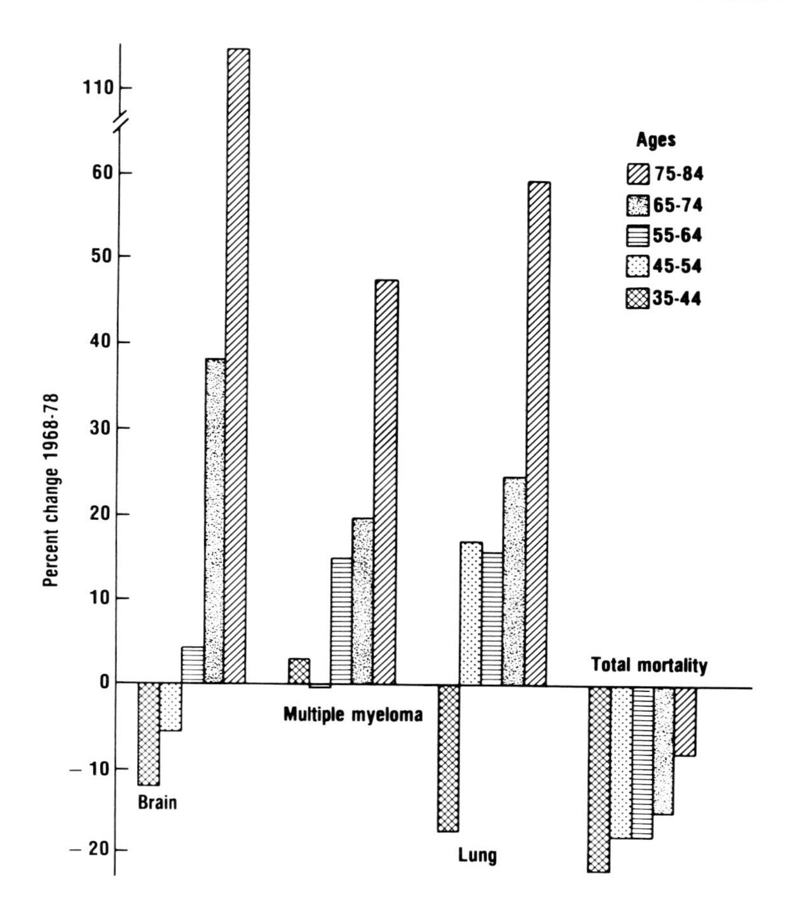

FIGURE 3. Cluster C: average percent change in brain and lung cancer and multiple myeloma and total mortality for men ages 35 to 84 from 1968 to 1978.

gistrar General's Decennial Supplement for England and Wales, 1970—72, Ser. D5, No. 1). Figure 4 is a re-drawing of Figure 6.4 from that volume. Crudely, social class 1 consists of professionals, social class 2 of "intermediate occupations", social class 3N of nonmanual skilled occupations, social class 3M of manual skilled occupations, social class 4 of partly skilled occupations, and social class 5 of unskilled occupations. The income data were derived from census data and may not be strictly accurate, particularly for social class 1. As Aaron Wildavsky said, "richer is safer" (at least the rich and higher social classes live longer). An SMR of 100 is the national average. Greater than 100 implies excessive deaths; less than 100 implies the opposite.

What Figure 4 shows is that social class is a great determinant of life and death as is income, and that they are highly correlated. Occupation often determines social class, and the greatest industrial and environmental exposures are likely to occur to persons in the lowest social class. Since class, i.e., jobs and income often determine lifestyle, it is difficult (for me, at least) to separate the two. For me, the word "lifestyle" is vague and diffuse. I can identify at least seven different lifestyles that I have lived over the last 60 years. If I were to develop a life-shortening disease, I am sure that some self-styled life stylists could blame at least one of the seven lifestyles, or possibly all of them. They may be right, but like choosing one's ancestors, there is not much to be done to change past lifestyles. If we are to prevent disease tomorrow, we should look to what we can do today.

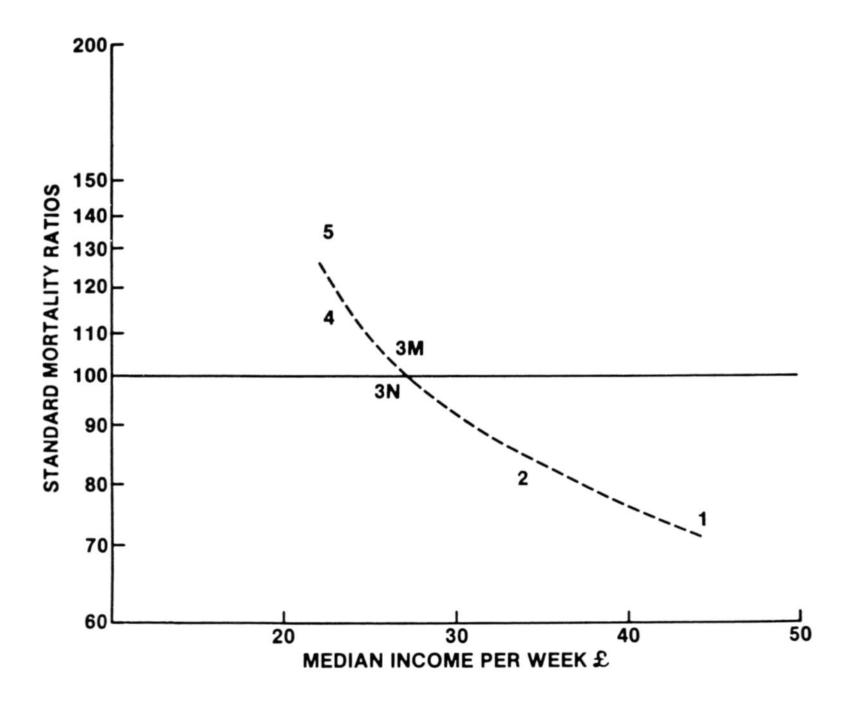

FIGURE 4. Standard mortality ratios by social class (1 to 5) and income, Great Britain: men, 1970 to 1971.

If we are to reduce life shortening, we need to consider what are reasonable goals. Figure 5 shows the probability for a male alive at age 15 entering any one of several occupations of surviving to age 65. Teachers, managers, and public officials have the highest probability of surviving (about 70%) to age 65. That 70% impresses me as a desirable goal for the entire work force. It is probably even achievable.

V. SUMMARY AND CONCLUSION

There is a real problem of cancers derived from, or associated with exposures at work or in the products (such as toxic wastes) that are consequences of waste. From a scientific point of view, there are great problems in clearly uncovering these hazards. Even the best of test systems employing the best new biology will leave us with many false positives and false negatives. Nonetheless, associating a small percentage of all cancers in the total population with industrial exposures means that a large percentage of cancers in exposed persons derives from their industrial exposures. I deliberately avoid the word "cause" because if cancer is a multistage disease, as most believe it is, to speak of a "cause" as if there were no important interactions with other "causes" impresses me as operationally misleading.

Trends in cancer mortality among men in the U.S. indicate that it is most unlikely that the increases seen since 1960 can be explained solely by cigarette smoking, especially since the proportion of adults smoking, the tar content of cigarettes smoked, and the number of cigarettes made per pound of tobacco have all declined since the early 1960s. What is more likely is that cigarette smoking (as the most important single contributor to cancer in the developed countries of the world) has contributed to the increase, but so too, have industrial exposures. The place of lifestyle in these changes is as hard to pin down as the definition of lifestyle. Social class, which is largely determined by occupation, is directly correlated with health, and inversely with death. A possible goal is to have all working people reach

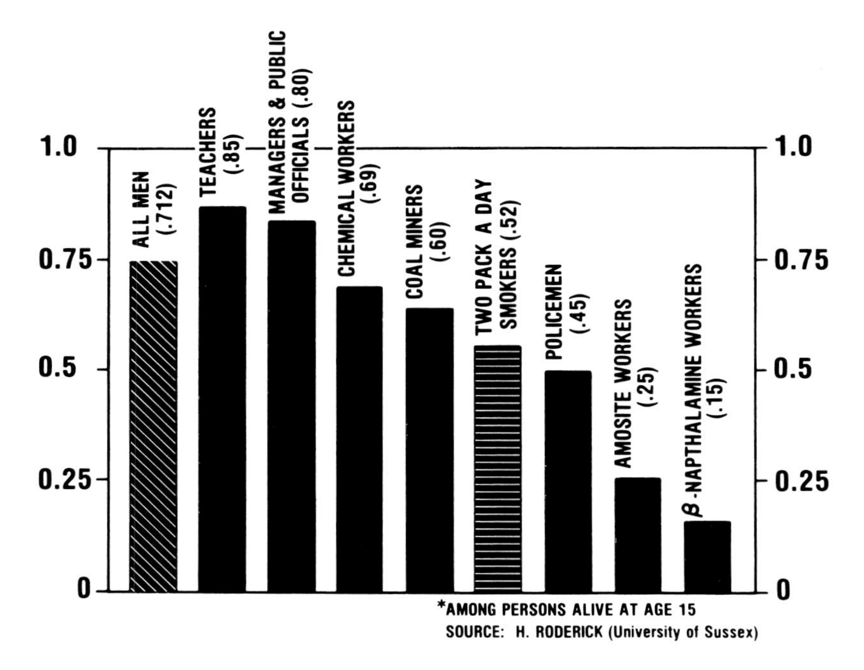

FIGURE 5. Probability of living to age 65 of white males in England and Wales (1971): some sample occupations.

the same probability of surviving to age 65 as do teachers, managers, and public officials. Resolving the argument about whether 1 or 5 or 15% etc. of all cancers are "due to" industrial exposure impressed Rosenstock[2] as not having much potential for improving our lives and our life expectation, a conclusion with which the author agrees.

From a social policy or public health point of view, there have been three ways of preventing disease, and thus increasing lifespan. All of them are necessary. The order in which they have been historically most successful is (1) things done by others, i.e., community action: good water supplies, effective waste disposal; (2) and (3) things we do ourselves: once (or rarely) in a lifetime, e.g., some innoculations, smallpox vaccinations, etc., and everyday, e.g., control our diets, our personal behavior, our smoking, our drinking, and our sex lives. The first of these three is the easiest to do. The last is the hardest.

Each of us is responsible for himself. Each of us also, by living in a symbiotic community, is his brother's keeper, i.e., has responsibility for others. The glass beer bottle that is discarded on the side of the road will cut a neighbor's foot. Both individual and corporate/community should behave responsibly in order to prevent disease. This refers to the original question: why did God make only one Adam? I think the answer we give has great implications for how we behave.

ACKNOWLEDGMENTS

I am indebted to many colleagues for the discussions that led to this paper — David Cruess, Devra Lee Davis, Vasilios Frankos, Jenice Longfield, Mary Jean Marvin, Mary Paxton, James Schlesselman, Carl Schulz, David Siegel, Jeffrey Trauberman, and Harry Wetzler, among others.

Sergeant First Class Michael Callahan (U.S.) gave considerable help with the figures. Mrs. Martha Ross managed this manuscript in all stages of development.

REFERENCES

1. **Peto, R. and Schneiderman, M., Eds.,** *Quantification of Occupational Cancer,* Banbury Report 9, Cold Spring Harbor Laboratory, Cold Spring Harbor, N.Y., 1981.
2. **Rosenstock, L.,** Review of Peto, R. 1981, *Med. Care,* 21, 471, 1982.
3. NAS/NRC, *Risk Assessment in the Federal Government: Managing the Process,* National Academy Press, Washington, D.C., 1983.
4. **Karstadt, M., Bobal, R., and Selikoff, I.,** A survey of availability of epidemiologic data on humans exposed to animal carcinogens, in *Quantification of Occupational Cancer,* Peto, R. and Schneiderman, M., Eds., Banbury Report 9, Cold Spring Harbor Laboratory, Cold Spring Harbor, N.Y., 1981.
5. **Acheson, E. D.,** Record linkage and the identification of long-term environmental hazards, *Proc. R. Soc. London Ser. B,* 205, 165, 1979.
6. **Beebe, G.,** Record linkage and needed improvements in existing data resources, in *Quantification of Occupational Cancer,* Peto, R. and Schneiderman, M., Eds., Banbury Report 9, Cold Spring Harbor Laboratory, Cold Spring Harbor, N.Y., 1981, 661.
7. **Smith, M.,** Long-term medicine follow-up in Canada, in *Quantification of Occupational Cancer,* Peto, R. and Schneiderman, M., Eds., Banbury Report 9, Cold Spring Harbor Laboratory, Cold Spring Harbor, N.Y., 1981, 675.
8. **Howe, G. R. and Lindsay, J. P.,** A follow-up study of a 10 percent sample of the Canadian labor force. I. Cancer mortality in males 1965—73, *J. Natl. Cancer Inst.,* 70, 37, 1983.
9. **Siemiatycki, J. et al.,** Discovering carcinogens in the occupational environment, *J. Natl. Cancer Inst.,* 66, 217, 1981.
10. **Siemiatycki, J. et al.,** Exposure-based case-control approach to discovering occupational carcinogens: preliminary findings, in *Quantification of Occupational Cancer,* Peto, R. and Schneiderman, M., Eds., Banbury Report 9, Cold Spring Harbor Laboratory, Cold Spring Harbor, N.Y., 1981, 471.
11. **Siemiatycki, J. et al.,** Preliminary report of an exposure-based case-control monitoring system for discovering occupational carcinogens, *Teratogen. Carcinogen. Mutagen.,* 2, 169, 1982.
12. **Doll, R. and Hill, A. B.,** The mortality of doctors in relation to their smoking habits: a preliminary report, *Br. Med. J.,* 1, 1451, 1954.
13. **Hammond, E. C. and Horn, D.,** Smoking and death rates: report on 44 months of follow-up on 187,783 men. I. Total mortality. II. Death rates by cause, *JAMA,* 166, 1159 and 1294, 1958.
14. **Dorn, H.,** The mortality of smokers and non-smokers, in *Proc. Soc. Stat. Sect. American Statistical Association,* American Statistical Association, Washington, D.C., 1959, 34.
15. **Hoar, S.,** Epidemiology and occupational classification systems, in *Quantification of Occupational Cancer,* Peto, R. and Schneiderman, M., Eds., Banbury Report 9, Cold Spring Harbor Laboratory, Cold Spring Harbor, N.Y., 1981, 455.
16. **Bridbord, K. et al.,** Estimates of the fraction of cancer in the United States related to occupational factors, in *Quantification of Occupational Cancer,* Peto, R. and Schneiderman, M., Eds., Banbury Report 9, Cold Spring Harbor Laboratory, Cold Spring Harbor, N.Y., 1981, 701.
17. **Horbach, L. and Loskant, H.,** Berufkrebstudies (A study of occupational cancer), in *Deutsche Forschungsgemeinschaft,* Harold Boldt Verlag, Boppard, Germany, 1981.
18. **Davis, D. L. and Magee, B.,** Cancer and industrial chemical production, *Science,* 206, 1356, 1979.
19. **Davis, D. L. et al.,** Estimating cancer causes: problems in methodology, production and trends, in *Quantification of Occupational Cancer,* Peto, R. and Schneiderman, M., Eds., Banbury Report 9, Cold Spring Harbor Laboratory, Cold Spring Harbor, N.Y., 1981, 285.
20. **Hoerger, F.,** Indicators of exposure trends, in *Quantification of Occupational Cancer,* Peto, R. and Schneiderman, M., Eds., Banbury Report 9, Cold Spring Harbor Laboratory, Cold Spring Harbor, N.Y., 1981, 435.
21. **Schneiderman, M.,** Trends in cancer mortality and incidence in the United States: is the future clear or clouded? in *Statistics in Medical Research,* Mike, V. and Stanley, K. E., Eds., John Wiley & Sons, New York, 1982, 71.
22. **Doll, R. and Peto, R.,** The causes of cancer: quantitative estimates of avoidable risks of cancer in the United States today, *J. Natl. Cancer Inst.,* 66, 1191, 1981.
23. **Davis, D. L. et al.,** Cancer prevention: assessing causes, exposures and recent trends in mortality for U.S. males, 1968—1978, *Teratogen. Carcinogen. Mutagen.,* 2, 105, 1982.
24. **Higginson, J. and Muir, C. S.** Environmental carcinogens: misconceptions and limitations to cancer control, *J. Natl. Cancer Inst.,* 63, 1291, 1979.
25. **Wynder, E. L. and Gori, G. B.,** Contribution of the environment to cancer: an epidemiological exercise, *J. Natl. Cancer Inst.,* 58, 825, 1977.

Chapter 2

CANCER AS AN OCCUPATIONAL HAZARD: EPIDEMIOLOGICAL EVIDENCE

Anthony B. Miller

TABLE OF CONTENTS

I. INTRODUCTION

This paper will be largely concerned with the use of epidemiology as a means to derive evidence of occupational hazards. Specific hazards will be mentioned only insofar as they relate to examples of the use of the epidemiologic method and will tend to be drawn from the experience of our unit in Toronto, either in developing methods or evaluating established methods in the occupational setting. The basic underlying thesis is that cancer as an occupational hazard in man can only be recognized by the application of the epidemiological method. This, in the light of the long latent period of cancer induction, suffers from the major disadvantage that a hazard must occur in sufficient numbers of individuals to be recognized before action can be taken. For this reason there are increasing endeavors to develop approaches to ensure that potential hazards are not introduced either into the occupational or the general environment. Many of these will be successful but we have the major problem that there are probably a number of substances that have been in use, possibly for decades, which may result in human disease which has so far not been recognized. Even if short-term and other tests applied in advance to new chemicals eventually prove to be entirely successful, because of the enormous number of chemicals currently in use which have never been tested, there will be a period during which surveillance by the epidemiologic method could be the only mechanism whereby previously unsuspected hazards are recognized. My intent is to indicate how these approaches can be refined and utilized in the most efficient and economical way possible.

II. DESCRIPTIVE STUDIES

Epidemiology is an observational science and at the simplest level describes what is going on in terms of the incidence or mortality of specific diseases in specific areas. An example of this approach is the mapping of areas with rates computed at the county or census district level. An atlas of cancer mortality covering the period 1950 to 1969 has in the U.S. been a fruitful source of hypotheses suggesting occupational hazards up until then unsuspected.[1] Examples include the various concentrations of bladder cancer in males in the industrial northeast, presumably largely explicable on the basis of chemical exposure, and the clusters of counties in coastal areas in the southeast with high rates of lung cancer now attributed to asbestos exposure during shipbuilding in World War II. Similar mapping has since been attempted in Canada, the U.K., and China. In Canada, mortality in the period 1966 through 1976 was studied.[2] The resulting maps have not been as informative as those in the U.S. There was as the expected concentration of lung cancer in urban areas, which we suspect is largely due to higher levels of cigarette smoking in urban than rural areas, though the expected excess attributable to fluospar mining in one area of Newfoundland is recognizable. Bladder cancer has hardly shown any excess, however, this does not mean, as I shall demonstrate later, that we do not have a problem with occupationally induced bladder cancer. One possible reason why the Canadian maps are not as informative as in the U.S. is the smaller population, often spread over larger areas. A second reason is the fact that mortality was aggregated only over 11 years, whereas a 20-year period was used in the U.S. On the other hand, it is also possible that Canada has less of a geographically based problem than in the U.S.

Descriptive studies can only point to a potential problem and even if one can attempt some correlation with the proportion of the work force employed in specific known occupations, this can only give rise to hypotheses and not test them.[3] More often than not, nonspecific excesses are found and often it may well prove that differences in socioeconomic status or differences in smoking rates may explain most of what is seen. The importance of socioeconomic status was well seen by studies based on the Registrar General's decennial supplement for 1970 to 1972 in England and Wales.[4] For many occupations, social class standardization resulted in an excess or deficit in a standardized mortality ratio, coming much closer to the norm and losing statistical significance. This happened for example, to mortality from malignant neoplasms in men aged 15 to 64, among miners and quarrymen, gas, coke and chemical workers, glass and ceramic workers, electrical and electronic workers, engineering and allied trades, woodworkers and warehousemen, and storekeepers. However, excesses were still noted in furnace, forge and foundry workers, food, drink and tobacco workers, construction workers, painters and decorators, and transport and communications workers. Sales workers which had a deficit in mortality reverted to normal with social class standardization. Professional and technical workers also moved toward normal though they still had a lower mortality than the general population. Based on these studies Fox and Adelstein evaluated the variation in mortality between various work orders associated with work or lifestyle. They concluded that lifestyle or socioeconomic status or other associated factors accounted for 88% of the variation of cancer rates and 82% of that associated with death from all causes.[5] This meant that only 12% of the variation in cancer and 18% for all causes could be attributed to factors associated with work.

III. ANALYTICAL STUDIES

Evidence confirming the importance of occupational factors in causing cancer has been largely derived from an initial observation of a cluster of cases by an alert physician, sometimes in occupational medicine, and subsequent confirmation of risk by a special analytical epidemiological study. The types of analytical epidemiological studies performed largely up to the present have been what have been variously called historical cohort studies or retrospective cohort studies. In this design, advantage is taken of records of individuals who have been employed in a plant or an industry in the past and their experience is ascertained by following them through to the present day or the recent past.

Although such studies have been responsible for identifying a number of hazards and still have a place in the epidemiological armamentarium, it is important to recognize that they have a number of difficulties and limitations. The first relates to the nature of the exposure information that is likely to be available. Pre-existing records rarely permit quantification of the amount of exposure to a particular substance or agent encountered by workers. Frequently exposure records just do not exist. Where they do, they suffer from the disadvantage of being averaged over work areas and rarely relate to individuals. Even where they did exist, there may be difficulties. Thus, radiation badge records may be missing, or they may be believed to be inaccurate. For example, in one group we are involved in, the early records were destroyed by fire. In some instances, it is known that workers who were reaching the maximum permitted exposure levels continued to work but omitted wearing their radiation dosimeter badge when they were in the dangerous areas. A number of attempts have been made to substitute for the lack of exposure information by looking at risk in relation to specific job titles. Job titles however, may not reflect the exposure of interest, and indeed it may be extremely difficult to know whether workers in a particular type of occupation categorized by a certain job title have or have not been exposed to the substance of interest. What is more, it may not be certain what constitutes the exposure of interest. This may occur particularly under circumstances where people have been exposed to multiple chemicals either in the same occupation or in a succession of different occupations. A further difficulty is that studies of this type rarely provide information on prior occupation before entering the specific plant or industry or subsequent occupations that an individual has had after leaving the plant. Yet their disease experience may be more related to these other occupations than those in the plant of interest. Finally, very few of the studies of this type have information on lifestyle factors that may confound the association being investigated. A classic difficulty is the absence of information on smoking in historical cohort studies devoted to lung cancer. There are other factors that may confound though information may not be available on them. These include diet, and those various undefined factors that fall under the heading ''socioeconomic status'' may yet influence cancer risk in a way that at times we cannot determine.

A second set of difficulties with historical cohort studies relates to the follow-up mechanism. Individuals may be lost after leaving a plant, they may migrate out of the area — sometimes out of the country, or they may change their name for a number of reasons including marriage. This may make it extremely difficult to trace the full cohort, yet it is important to ensure that the total group exposed are followed as the workers who move may have been a particularly high risk group, and their experience may be critical in determining the extent of any possible hazard. Some studies, including one of our own, have been based on pensioners. Herein is a considerable danger of selection. In general there are minimal criteria applied before someone can become a pensioner. They may have to have worked for 5 or 10 years. Sometimes they have continued working until the time they are normally due to retire in their 60s, so that once again such groups are unrepresentative of the general occupational labor force. Very often pensioners have been healthier than the general population. Someone who dies while working is not likely to become a pensioner. An early hazard that might be fairly early identified in a workforce which is fully followed from the time of initiation of employment may be missed or masked in studies devoted to pensioners.

The other difficulty with follow-up in historical cohort studies relates to the danger of ascertainment bias. Special surveillance increasingly is applied to certain groups suspected of being at risk of a particular hazard. Examples include periodic X-rays for those at risk of asbestosis and cancer of the lung, sputum cytology for similar groups and for radiation workers, and regular medical examinations for a number of different occupational groups. The mere process of surveillance may thus bring to light disease, including cancer, that might otherwise not have been detected. Ascertainment bias of this type may be magnified if a study involves special evaluation of, for example, diagnostic records, hospital records, or even of death certificates by the investigator. In general, evaluation of this type cannot be performed on the data that may be compiled for the control group for the study, if it is derived, e.g., from the general population. Procedures of this type applied to one group but not to the comparison group are a particular example of bias that should be avoided in a historical cohort study.

The third set of difficulties with historical cohort studies comes at the stage of analysis and interpretation. A major difficulty may be the lack of an adequate control group. Many historical cohort studies may be based on a comparison with the mortality expected in a group of the same age and sex distribution and similar time period using data from the same province or state or country. A well–recognized problem with such comparisons is what has been called the healthy worker effect. A number of deaths in people of working ages occur among individuals who may no longer be in the work force because of disease or disability. Workers in general are healthier than the general population. They are more active, and the process of job selection will tend to weed out those with disabilities. Thus, their mortality from many causes, especially from heart disease is likely to be less than the general population. Even cancer mortality is likely to be less in the early stages of the observation of a cohort because individuals with cancer are usually not present at the time a cohort is established for subsequent observation. Although for some cancers the interval between diagnosis and death is short, for many the period will exceed 2, sometimes 5 years. Thus, if one establishes a cohort in a particular year, for example 1950, for the first few years post-1950 one can expect a low mortality from cancer. In general, for cancer the healthy worker effect tends to disappear after a 5- and certainly a 10-year period, but its influence persists for other causes, especially cardiovascular disease. This has a major effect on the type of analysis that might be performed. There is a temptation to look at deaths only and to compare their distributions with the distribution derived from the comparison group, usually the general population. If there is a deficiency of cardiovascular disease deaths because the workers are healthier than the general population, then it is easy to see how there could be a relative excess of cancers. Proportional mortality ratios simply look at the proportion of deaths due to cancer in comparison with that expected in the general population. They are therefore particularly affected by the healthy worker effect and cancer excesses which are spurious may be noted in such studies. Unless comparisons can be made with other employed groups, such numerator type studies should be avoided and major efforts made to ensure that the denominator of those at risk from cancer is adequately identified. The expected numbers, based on mortality from the comparison population, should be derived in an appropriate way. Hence, for historical cohort studies standardized mortality ratios are much preferred to proportional mortality ratios.

A further difficulty however, with interpretation of historical cohort studies is the way that exposure can be correlated with duration of employment and survival. Generally, this effect occurs because exposure is to a relatively weak carcinogen and its effect is only detectable after a certain dose has been accumulated. Thus, in most circumstances there must be a substantial period of employment before an effect will be noted, while the greater the period of employment the greater the exposure and the greater the probability of an effect occurring. On the other hand, it is possible that those who have more intense exposures may have a shorter survival than those who have a lower level of exposure because of development of disease related to the exposure. One may then encounter the paradox that individuals with short durations of exposure will have higher rates of disease than those with long durations of exposure. Yet, if exposure records are not available, such a finding may be almost impossible to evaluate. A more extreme example of this could occur if the only individuals who contract disease are those who are particularly susceptible. If susceptibility is manifested by other symptoms or the early development of disease, they may be removed from exposure and the disease may only be seen in those with short periods of exposure and not those with long periods of exposure. This could be marked by an inverse correlation between duration of exposure and development of disease or survival. Further problems may occur in relation to analyses of duration of employment. There may be a tendency in studies to compare individuals with similar durations of employment. This could be a direct correlate with the exposure variable and the effect may be matched out. In addition, some investigators have included the experience of individuals prior to the date of establishment of a cohort in their analysis and pointed to low rates of disease in the early years. Clearly however, if the study only includes the survivors after a period following initiation of exposure, it may show no disease occurring prior to the date of establishment of a cohort.[6-9]

Yet another difficulty associating exposure to duration of employment relates to whether one includes in the study individuals with short durations of exposure and those who have left the industry. These people should be included if at all possible because they may include a group whose exposure has been short, either because they exhibited symptoms and may indeed be susceptible to the agent, or because the exposure was so intense that they decided they could no longer permit themselves to be exposed to the potential risk. To exclude individuals with short durations of exposure and particularly those who may have left the company, may therefore mask the real extent of risk.

A final difficulty with historical cohort studies relates to the fact that they are usually conducted on groups in whom one has a prior evidence of a hazard. This hazard may have become noted because this particular group was heavily exposed. There is a temptation to regard findings from such studies as being generally applicable. However, if they had been derived from groups that are heavily exposed, they may not be generalizable and not represent the problem which exists in the total population exposed to this particular hazard. Workers who had been exposed to low dose levels may not have been studied at all, or may have been relatively few (at least in terms of the numbers required to demonstrate an increase in disease risk) so that precise estimates of risk at low dose levels may not have been possible. Hence, attempts to determine dose-response relationships are likely to be crude and very imprecise. Some rather unrealistic estimates that were widely circulated a few years ago over the extent of cancer attributable to occupation seem largely to have been developed because of failure to appreciate this difficulty.[10]

Some of the difficulties with historical cohort studies can be overcome by prospective cohort studies, i.e., those established at present with follow-up into the future. Such studies suffer from the disadvantage of having to wait a long time to demonstrate an effect. However, they need not be very expensive, as information on prior and current exposure and on lifestyle factors can be collected relatively easily on an existing work force in some detail. Then the work force can be subsequently monitored by means of routine approaches. Such monitoring should clearly include determination of changes in exposure levels.

IV. CASE-CONTROL STUDIES

An alternative approach is to utilize the case-control methodology, which has probably not been sufficiently applied to evaluating the extent of occupational risks. Case-control studies are in general disease-specific. Thus, they would be applied to individuals with lung cancer, bladder cancer, etc., although on occasion case-control studies designed to assess occupational exposures have been applied to multiple sites. Nevertheless, single site case-control studies are not necessarily a disadvantage as very often they would be conducted under circumstances where they are seeking to confirm a hypothesis that a specific disease is increased by a certain exposure derived from another investigation.

One of the major advantages of case-control studies is that multiple exposures are ascertained and lifetime occupational histories are possible. Thus, within the limitations of recall by individuals of what they have done, it is possible to evaluate the effect of confounders in relation to specific exposures. One is also not entirely dependent upon an individual's memory in terms of their occupation. With their permission, questions can be addressed to appropriate occupational settings and further details of work and potential exposures ascertained. This is particularly the case when a case-control study is being conducted within a specific industry or plant, but it is also possible when a case-control study is being performed in a region where there is suspicion of a specific occupational risk.

One possible extension of the case-control methodology to determine the extent to which exposures can be specifically identified from occupational histories is being evaluated by Siemiatycki in Montreal.[11] An occupation and exposure linkage system for the study of occupational carcinogenesis has also been developed in Boston.[12] However, this latter system has not yet proven generalizable to the exposures encountered in our own studies. Further work along these lines, therefore, is highly desirable.

A major advantage of case-control studies is that the attributable risk from a particular exposure can be determined. A study conducted within a geographically defined area in which all incident cases are ascertained over a specific time period with controls drawn from the general population, enables one to quantify the extent to which the disease can be attributed to a specific exposure. A particular example of this was a case-control study which we performed in the Toronto area after it was suggested in another study that a high proportion of cases of laryngeal cancer were attributable to asbestos exposure.[13] By ascertaining a population-based series of cases and controls we were able to demonstrate that in this area most of the disease was attributable to tobacco and alcohol use. On an individual basis only

3% of the cases were attributable to asbestos exposure but the asbestos exposure did not add to the amount of disease that could be explained from exposure to tobacco and alcohol. Thus, in spite of the previous study, it is clear that asbestos and for that matter other occupational exposures are only responsible for a minority of cases of laryngeal cancer in our area. Nevertheless, they are important in some instances and this study has helped to confirm the validity of providing compensation for cases of laryngeal cancer with occupational exposure to asbestos with the disease occurring after an appropriate interval following initiation of exposure.

Cole,[14] in Boston, was one of the first to use a case-control study to quantify the amount of a disease attributable to occupation. In this study, an increase in risk of lower urinary tract cancer from exposures to dyestuffs, rubber, rubber products, leather, leather products, and other organic chemicals was found. The attributable risk percent from hazardous occupations was 18 in males and 6 in females. Our first study of bladder cancer was conducted primarily in areas where we had not anticipated much in the way of occupationally induced bladder cancer, namely British Columbia, Newfoundland, and Nova Scotia.[15] Nevertheless, we found that 25% of bladder cancer in males and 6% in females was attributable to industrial exposure. Many of those who developed occupationally related bladder cancer appeared to have migrated to the area and developed cancer after retirement. Looked at on an individual occupational basis, it was clear that much of this risk was due to exposure in *a priori* suspect industries: chemical, rubber, photographic, petroleum, medicine, and food processing. Nevertheless, a general search of the occupational history revealed other occupations which suggested increased risk including guards and watchmen, nurserymen, metal machinists, material recorders, members of the armed services, glass processors, mechanics, and clerical workers. The presence of an increased risk in guards and watchmen probably reflects prior exposure in a different occupation before taking up such an occupation relatively late in life. Another type of exposure which increased risk was exposure to dust or fumes. Within this we found increased risk for those exposed to crop spraying, welding, railroad working, and diesel and traffic fumes.

The possible risk in railway workers has been evaluated through the Canadian National Railways pension file.[16] Of 43,826 individuals who were on the file during the period 1965 to 1977, 5.8% had retired prior to 1950, 21.4% from 1950 to 1959, 36.6% from 1960 to 1969 and 36.2% from 1970 to 1977. Although fact of death was recorded on the file, cause of death was not. The file was linked with the National Death Index to determine the cause of death for those 17,838 individuals who were known to have died during 1965 to 1977. In fact, no excess of bladder cancer was found. There was, however, a possible excess of other urinary cancer and of lung cancer but not of all cancers or all causes of death. The increased risk of lung cancer was explored further, and it was found to be probably associated with exposure to diesel fumes and/or coal dust. This is an observation previously cited and may justify further study. Although I indicated before that retirees may not be a good group to look into for increases in risk because of the possibility that the susceptibles may develop disease prior to their timed retirement, this may not be true for bladder cancer which often, even if occupationally induced occurs later in life. However, it is apparent that the hypothesis proposed in a case-control study was not confirmed in the investigation subsequently performed and this must remain a hypothesis subject to further evaluation. Nevertheless, the potential association of railway workers with lung cancer is one that possibly also needs to be pursued and we are now conducting a case-control study of lung cancer, which even though primarily directed to women will provide some information on the extent to which lung cancer is occupationally induced in men. It will be of interest to see if the association with diesel fumes is noted again in that study.

V. EPIDEMIOLOGIC MONITORING

The CNR study just described was an example of a study based on computerized record linkage. I should now like to describe another study based on computerized record linkage, conducted by Dr. G. R. Howe of the NCIC Epidemiology Unit, which indicates the way in which computerized record linkage, using available records, can provide information relevant to the monitoring of the general population for potential occupational hazards.[17] This study is based on the book renewal file of the Canadian Social Insurance Number, which at the time the study was commenced included over 3 million records, and more specifically, an annual labor force survey of 10% of the occupational work force in Canada from 1965 to 1969 and a further sample of 5% of the labor force in 1971, together comprising records of 700,000 individuals. The labor force surveys were based on the terminal digit of the social insurance number and therefore largely consist of information on the same individuals repetitively sampled over 5 years, and then one half of them 2 years later. These data were linked to the Canadian Mortality Data Base which currently includes records of all deaths in Canada in a computerized linkage form from 1950 through 1980. At the time the first linkage was performed, the period 1965 through 1973 was covered with an output of approximately 30,000 deaths. Thus, occupational information is available on a basic cohort of 700,000 individuals, 30,000 of whom died in the period 1965 through 1973.

The process of record linkage involves a number of steps. First, the data has to be made consistent; names have to be coded into a linkable form which eliminates errors due to change of spelling. The Canadian system uses an approach called Nysiis. Files are then sorted into large blocks of a size suitable for record linkage with the output file. The linkage is run and weights are calculated in terms of the probability that the link is in fact a true one. A decision is made each time a linkage is run on the appropriate cut off point for the weight. If the cut off point is too low then a number of false positive links will be included in the analysis; if too high, a number of true positive links will be eliminated. It is particularly important in record linkage studies that both files being linked include adequate personal identifiers.

This has been largely achieved for the mortality data base in Statistics Canada but it is important to note the identifiers that should be present in any file which is likely to be linked with it. Critical identifiers are the surname and for women the maiden name, or for both sexes other name change(s) during life, first names rather than initials, and wherever possible mother's maiden name, all helping to categorize an individual as unique. Marital status for women is informative and sex, though sometimes derivable from the name should generally also be included in the file. Most critical is full date of birth — day, month and year, and birthplace — city and province if born within Canada, country if place of birth is outside Canada. If unique personal identification numbers are available, a great deal of these personal identifiers are not required, but are still desirable because of errors in numbering systems. In Canada, there is no universal unique personal identifier. In some studies however, the Social Insurance Number applicable to employed persons can be used; in other circumstances, the Health Insurance Number, but at present neither are available on death records. Nevertheless, for an occupational group where social insurance numbers would generally be available from employment records, a link with the Social Insurance Number book renewal file would provide the other identifiers if they are deficient on the employment records.

For this particular study, a series of steps were necessary. The occupational records for individual years were sorted by Social Insurance Number to form a composite record. The composite occupational records by linkage with the Social Insurance Number master index file then had identification data added to them. This provided occupational records individually identified for linkage. Linkage was then made with the mortality data base to provide the occupation death links. Linkages are planned with cancer incidence data held nationally by Statistics Canada to provide occupational cancer incidence links.

So far the study has not been used for monitoring though there are some interesting findings. Looking at occupations, bartenders and waiters have a high SMR for buccal cavity and pharynx cancer. Bartenders also have high SMRs for cancer of the trachea, bronchus, and lung. Stomach cancer is found in excess in farm laborers, linemen and servicemen, and metal workers. Looking at various industries, lung cancer is found in excess in the bowling alley and billiard parlor industry and in machine shops. The mortality link has now been extended to include deaths up to 1979. It is hoped that further leads will follow from this. At the moment however, it seems clear that some of the excesses are to be expected from lifestyle exposures, particularly an excess of lung cancer in occupations and industries where it is known that there is a great deal of cigarette smoking. Nevertheless, this study is a prime example of the utility of record linkage studies using routinely collected data. It will provide a monitoring hypothesis deriving system for years. It also provides the means to test new hypotheses.

One question that arises in the mind of the public as well as scientists is the problem of confidentiality in record linkage projects. A number of safeguards enable us to ensure that confidentiality is preserved. Most important is the integrity of the researcher. However, all such studies should only be undertaken following human experimentation committee review, and wherever possible the consent of the subjects should be obtained. Although not possible for the study I have just described, in a current study in which we expect to use record linkage to determine endpoints (the National Breast Screening Study in Canada), it is possible to ensure that the consent forms include permission to link with vital records in the future.[18] However, the main precaution for confidentiality is security of records. Files containing identifying information are kept separate. In Canada the linkages with national mortality and cancer incidence files are conducted with full confidentiality within Statistics Canada, and linked records are not returned with identifying information on them to the investigator except under very special circumstances.

In order to design appropriate epidemiological surveillance it is always important that one should have an indication of whether risk is likely to follow from exposure to a particular substance. The various approaches available that would permit indirect assessment of risk in man are qualitative at best. Thus, although short-term tests for mutagenicity have been developed, we can question how predictive they are for carcinogenicity in animals, let alone for man. Nevertheless, there is sufficient correlation that two positive tests in different systems suffice to justify animal testing in the opinion of the Special Advisory Committee on Carcinogens of the Advisory Council on Occupational Health and Occupational Safety of the Ontario Ministry of Labour.[19] Short-term tests have now been accepted by the International Agency for Research on Cancer as part of the decision matrix as to whether or not a substance is in fact potentially carcinogenic to man.[20] For animal tests, we have problems of species sensitivity; therefore, to what extent is a test positive in a rat indicative of potential problems in man? Even more so there are problems of site specificity; therefore, is bladder cancer in a rat suggestive of bladder cancer in man? This may have been answered in the positive for saccharin but it is not necessarily also true for many other carcinogens. We also have the problem of extrapolation from animals to man. Can we use the apparent strength of a carcinogen in animals as an indication of apparent strength in man? Most people at present feel we cannot.

We are therefore faced with the question of how we can reduce the risk of occupational cancer. The Special Advisory Committee on Carcinogens in Ontario[19] proposed that all new chemicals have as a minimum two short-term tests, which if positive would require animal tests in two species. If the substance is an animal carcinogen, it should probably be banned. The difficulty arises in terms of all untested existing chemicals. Clearly, for known carcinogens, the aim should be to find a substitute and in the absence of this, aim for zero exposure of workers at risk. In the view of Doll and Peto,[21] laboratory data should be used solely for priority setting. However, in the view of a special committee established by the state of California,[22] such data are utilizable in terms of making risk benefit decisions.

Nevertheless, the most important thing to remember is that in spite of all the care with short-term and animal tests not all potential human carcinogens will be identified and thus, epidemiological monitoring is important with nonsuspected new chemicals for unusual occurrences, such as angiosarcoma of the liver attributable to vinyl chloride exposure, or for increases in the frequency of common disease. For known carcinogens there is a different reason for epidemiological surveillance. The object of surveillance in this instance is to provide protection for the individual at risk. This principle is already established for radiation workers for whom exposure records are maintained and in general detectors are used so as to monitor exposure up to a certain permissible annual level. If, as will often be the case, it proves impossible to prevent the occurrence of cancer, such monitoring enables the provision of compensation for disease if it occurs in spite of precautions taken. It is essential that research into appropriate prophylactics and secondary prevention continue in such exposed populations.[23]

Finally, we should consider the mechanism of epidemiological monitoring. We should ensure the maintenance of personal identifiers of those exposed to potentially carcinogenic substances. We should also require good records of exposure. This clearly is the responsibility of management with the collaboration of labor. Individuals at risk should be followed through plant, industry, and union records, or through specific identification in special registries. A disease risk should be identified not only by looking for unusual occurrences, but by planned record linkage, either with local registries or because of problems with migration, more appropriately with the national cancer reporting system and the National Death Index.

In the future we can look for considerably greater sophistication in the precision of epidemiological monitoring. Developments in molecular genetics may enable us to identify subsegments of exposed populations that may be particularly susceptible to the effects of carcinogens. Individual variation that may have a genetic basis, similar to the polymorphism noted in the handling of pharmacologic agents, is now well recognized.[24] Tools are being developed that will identify the footprint of carcinogenic interaction with DNA, the identification of DNA adducts.[25] Although such tools at present only identify continuing exposure, and thus may only have an immediate application in prospective cohort studies, other tools that could be used in case-control studies or historical cohort studies are undoubtedly around the corner.

VI. CONCLUSION

Epidemiology has been responsible for identifying a number of carcinogenic hazards in the occupational environment in the past and will be used for this purpose again in the future. It is important that we fully use the methods at our disposal: descriptive, historical cohort, especially designed prospective cohort, and appropriately designed case-control studies. In addition, the monitoring mechanisms that are now possible using computerized record linkage have to be developed and made available for this purpose. Many have objected to the potential expense of epidemiological monitoring, yet the records that facilitate this are available, and with modern computer systems can be put into an appropriate format suitable for record linkage purposes. The studies that have been performed have pointed the way and the lessons are already being put to use by responsible government agencies. Industry increasingly recognizes the need to put their own house in order and to ensure that their health systems are staffed by qualified epidemiologists. Those of us in academia must collaborate in this endeavor and ensure that the mechanisms that now exist are appropriately directed toward cancer control in the occupational as well as the general environment.

REFERENCES

1. **Mason, T. J., McKay, F. W., Hoover, R., Blot, W. G., and Fraumeni, J. F., Jr.,** Atlas of Cancer Mortality for U.S. Counties, 1950—1969 DHEW Publ. No. (NIH) 75-780, U.S. Dept. of Health, Education and Welfare, Washington, D.C., 1975.
2. Health and Welfare Canada, Statistics Cancer. Mortality Atlas of Canada Vol. 1, Cancer, Minister of Supply and Services Canada, Ottawa, 1980.
3. **Blot, W. J., Brinton, L. A., Fraumeni, J. F., Jr., and Stone, B. J.,** Cancer mortality in U.S. counties with petroleum industries, *Science*, 198, 51, 1977.
4. Office of Population Censuses and Surveys, Occupational Mortality, the Registrar General's Decennial Supplement for England and Wales 1970—72, Ser. D.S. No. 1, Her Majesty's Stationery Office, London, 1978.
5. **Fox, A. J. and Adelstein, A. M.,** Occupational mortality: work or way of life? *J. Epidemiol. Commun. Health*, 32, 73, 1978.
6. **Duck, B. W., Carter, J. T., and Coombes, E. J.,** Mortality study of workers in a polyvinyl-chloride production plant, *Lancet*, 2, 1197, 1975.
7. **Wagoner, J. K., Infante, P. F., and Saracci, R.,** Vinyl chloride and mortality? *Lancet*, 2, 194, 1976.
8. **Berry, G. and Rossitar, C. E.,** Letter: vinyl chloride and mortality? *Lancet*, 2, 416, 1976.
9. **Fox, A. J.,** Letter: vinyl chloride and mortality? *Lancet*, 2, 417, 1976.
10. **Bridbord, K., Decoufle, P., Fraumeni, J. F. et al.,** Estimates of the fraction of cancer in the United States related to occupational factors, unpublished manuscript, 1978.
11. **Siemiatycki, J., Day, N. E., Fabry, J., and Cooper, J. A.,** Discovering carcinogens in the occupational environment: a novel epidemiologic approach, *J. Natl. Cancer Inst.*, 66, 217, 1981.
12. **Hoar, S. K., Morrison, A. S., Cole, P., and Silverman, D. T.,** An occupation and exposure linkage system for the study of occupational carcinogenesis, *J. Occup. Med.*, 22, 722, 1980.
13. **Burch, J. D., Howe, G. R., Miller, A. B., and Semenciw, R.,** Tobacco, alcohol, asbestos and nickel in the etiology of cancer of the larynx: a case-control study, *J. Natl. Cancer Inst.*, 67, 1219, 1981.
14. **Cole, P.,** A population based study of bladder cancer, in *Host Environment Interactions in the Etiology of Cancer in Man*, Doll, R. and Vodopija, I., Eds., International Agency for Research on Cancer, Lyon, 1973, 83.
15. **Howe, G. R., Burch, J. D., Miller, A. B., Cook, G. M., Esteve, J., Morrison, B., Gordon, P., Chambers, L. W., Fodor, G., and Winsor, G. M.,** Tobacco use, occupation, coffee, various nutrients and bladder cancer, *J. Natl. Cancer Inst.*, 64, 701, 1980.
16. **Howe, G. R., Fraser, D., Lindsay, J., Presnal, B., and Yu, S. Z.,** Cancer mortality (1965—1977) in relation to diesel fume and coal exposure in a cohort of retired railway workers, *J. Natl. Cancer Inst.*, 70, 1015, 1983.
17. **Howe, G. R. and Lindsay, J. P.,** A follow-up study of a ten-percent sample of the Canadian Labor Force. I. Cancer mortality in males, 1965—73, *J. Natl. Cancer Inst.*, 70, 37, 1983.
18. **Miller, A. B., Howe, G. R., and Wall, C.,** The national study of breast cancer screening, *Clin. Invest. Med.*, 4, 227, 1981.
19. **Mustard, J. F., Farber, E., Miller, A. B., McCalla, D., and Till, J. E.,** Report of the Special Advisory Committee on Carcinogens, 4th Annu. Rep. Advisory Council on Occup. Health and Occup. Safety, Ontario Ministry of Labour, Toronto, 1982, 143.
20. International Agency for Research on Cancer, *IARC Monographs on the Evaluation of the Carcinogenic Risk of Chemicals to Humans*, IARC Monogr. Suppl. 4, IARC, Lyon, 1982, 10.
21. **Doll, R. and Peto, R.,** The causes of cancer. Quantitative estimates of avoidable risks of cancer in the United States today, *J. Natl. Cancer Inst.*, 66, 1191, 1981.
22. State of California Carcinogen Policy, Sect. 3: a policy for reducing the risk of cancer, Health and Welfare Agency, Department of Health Services, Sacramento, 1982.
23. **Magnus, K. and Miller, A. B.,** Controlled prophylactic trials in cancer, *J. Natl. Cancer Inst.*, 64, 693, 1980.
24. **Harris, C. C., Autrup, H., Connor, R., Barrett, L. A., McDowell, E. M., and Trump, B. F.,** Interindividual variation in binding of benzo [a] pyrene in DNA in cultured human bronchi, *Science*, 194, 1067, 1976.
25. **Perera, F. P. and Weinstein, I. B.,** Molecular epidemiology and carcinogen-DNA adduct detection: new approaches to studies of human cancer causation, *J. Chronic Dis.*, 35, 581, 1982.

Chapter 3

INTERACTION OF HOST AND LIFESTYLE FACTORS WITH OCCUPATIONAL CHEMICALS IN CANCER CAUSATION

Gary M. Williams

TABLE OF CONTENTS

I. INTRODUCTION

A number of industrial chemicals or processes have been shown to cause cancer among workers (Table 1). In these situations, fortunately, not every worker develops cancer. To a large extent, the degree of risk is determined by the extent of exposure.[1-5] However, there is clear evidence that other factors affect the outcome of exposures. This was in fact evident in the first demonstration of cancer causation in humans which was an occupational cancer. In 1775, Percival Pott described scrotal cancer among chimney sweeps in London. This cancer was apparently less common among similar workers on the Continent who observed better hygienic practices, thus resulting in lesser risk.[1]

Insight into the ways in which different factors contribute to the development of cancer has been provided by mechanistic studies in experimental chemical carcinogenesis. Extensive research has demonstrated that the induction of cancer in experimental animals is a complex process comprised of a number of distinct steps. These are part of two mechanistically separate sequences, the first being neoplastic conversion of the cell in which a normal cell is transformed into a neoplastic cell, and the second being neoplastic development and progression in which the neoplastic cell grows into an overt neoplasm. Considerable evidence indicates that the first sequence, which is also called initiation, is the result of an alteration in the genetic material of the cell. The second sequence, which includes the phenomenon of neoplasm promotion, results from the selective proliferation of the abnormal cell produced in the first sequence. Neoplastic development could also be facilitated by additional genetic effects, but more likely is produced by other actions. Thus, chemicals that produce neoplastic conversion of the cell generally have the ability to undergo chemical reaction with DNA and have been referred to as genotoxic. Those agents which operate to affect neoplastic development exert other biological actions, which have been categorized as epigenetic. For example, the process of promotion may result from effects on the cell membrane leading to selective cell proliferation.

The causation of many types of human cancer is clearly multifactorial,[6-9] resulting from the action of both genotoxic carcinogens and epigenetic enhancing agents.[9] Thus, in assessing the causes of occupational cancer, it is important to recognize that exposures in the workplace could contribute either the genotoxic agent that produces neoplastic conversion or the epigenetic factors that facilitate neoplastic development.

Several types of factors distinguished by their origins influence carcinogenesis (Table 2). Host factors are those that are inherent attributes of the individual, while the other factors are exogenous. Lifestyle factors are essentially those that are voluntarily acquired, whereas environmental factors are contributions of the environment. Occupational factors stem from specific environments and include both industrial chemicals and working conditions that lead to exposures to other hazardous agents, such as sunlight. These factors interact with one another in a variety of ways to determine cancer risk.

II. INFLUENCE OF HOST FACTORS ON OCCUPATIONAL CANCER

Among the occupational carcinogens shown in Table 1, there are several aromatic amines (4-aminobiphenyl, auramine, benzidine, and 2-naphthylamine) which have caused bladder cancer in workers.[10] The carcinogenicity of aromatic amines in experimental animals is determined by genetically controlled enzymes involved in chemical biotransformation,[11,12] and thus enzymatic differences in humans could translate into different levels of risk from exposure to this type of carcinogen.

The major route of aromatic amine biotransformation in most animal species is acetylation of the nitrogen atom in the amino group.[13] Humans display a hereditary polymorphism of

Table 1
INDUSTRIAL CHEMICALS OR PROCESSES FOR WHICH HUMAN EXPOSURE WAS EVALUATED AS CARCINOGENIC BY THE IARC[a]

Chemical	Type of cancer
4-Aminobiphenyl	Bladder
Arsenic and certain arsenic compounds	Skin
Asbestos	Mesothelioma
Manufacture of auramine	Bladder
Benzene	Leukemia
Benzidine	Bladder
Bis(chloromethyl)ether and technical-grade chloromethylmethyl ether	Lung
Chromium and certain chromium compounds	Lung
Industries	
Boot and shoe manufacture and repair (certain occupations)	Nasal
Furniture manufacture	Nasal
Rubber industry (certain occupations)	Bladder
Manufacture of isopropyl alcohol (strong-acid process)	Nasal, laryngeal
Mustard gas	Lung
2-Naphthylamine	Bladder
Nickel refining	Nasal
Soots, tars, and oils	Skin, lung, bladder, intestine
Underground hematite mining (with exposure to radon)	Lung
Vinyl chloride	Liver, brain, lung

[a] International Agency for Research on Cancer.[10]

Table 2
FACTORS AFFECTING HUMAN CANCER RISK

- Host
  - Genetics, sex
  - Age
- Lifestyle
  - Diet
  - Practices
  - Customs
- Environmental
  - Air & water contaminants
- Occupational
  - Chemical products
  - Working conditions

N-acetylation, such that individuals can be classified as either rapid or slow acetylators of a variety of aromatic amines. Slow acetylators are homozygous for a recessive allele and rapid acetylators are either homozygous or heterozygous for a dominant allele. In North American populations, the distribution of rapid and slow acetylators is approximately equal.

Evidence from experimental studies shows that, for carcinogenic aromatic amines, rapid acetylation leads to liver cancer, while slow acetylation leads to bladder cancer. Rabbits, like humans, are polymorphic acetylators, and it has been demonstrated that several aromatic amines are more genotoxic to hepatocytes with rapid acetylation capability than those with slow acetylation.[16,17] Such observations suggest that among humans occupationally exposed to aromatic amines, slow acetylators may be at a greater risk than rapid acetylators for bladder cancer, probably because of less activation in the liver. Several studies now show a slight excess of slow acetylators among bladder cancer patients with a history of occupational exposures.[16,17] Conversely, it could be that rapid acetylators are at greater risk for liver cancer, or at other sites of action of aromatic amines, such as breast.

III. INFLUENCE OF LIFESTYLE FACTORS ON OCCUPATIONAL CANCER

Lifestyle factors, particularly cigarette smoking, alcohol consumption, and diet play a major role in the etiology of the most common cancers in western industrialized countries.[18] These factors interact with one another to determine risk; for example, cigarette smoking and heavy alcohol consumption together lead to a high risk of cancer of the upper alimentary tract.[19,20] Thus it is to be expected that lifestyle factors would interact with occupational agents in disease causation.

Indeed, one of the most potent interactions in the enhancement of human carcinogenesis is the synergism resulting from cigarette smoking and exposure to asbestos.[21,22] In the absence of cigarette smoking, inhalation of asbestos fibers leads to the development predominantly of pleural mesotheliomas. This appears to be due to a physical effect, since other types of fibers, including glass fiber also produce experimental pleural neoplasms. Moreover, asbestos has not been shown to be genotoxic.[23] Exposure to both asbestos and cigarette smoking greatly increases the occurrence above that produced by cigarette smoking alone, of the main type of cancer resulting from smoking, namely, bronchogenic carcinoma.

The combined effect of asbestos and cigarette smoking could result from several mechanistically distinct interactions.[22] Support for a cocarcinogenic effect of asbestos, which would be characterized by an increase in neoplastic conversion of cells, has been provided by cell culture studies in which the concurrent application of asbestos and polycyclic aromatic hydrocarbons (PAH) of the type that occur in cigarette smoke leads to increased mutagenesis[24] and transformation.[25] These types of enhancement could occur through facilitation by asbestos of the uptake of the carcinogen into cells.

The chrysotile form of asbestos has also been shown to exert a promoting effect on the development of neoplasms in tracheal transplants previously exposed to PAH.[26] Asbestos is cytotoxic to a variety of cell types,[27,28] and it could be that in the lung, killing of cells stimulates compensatory hyperplasia, resulting in the enhancement of neoplastic development. Thus the synergistic effects of asbestos and cigarette smoking in humans could occur as a consequence of both cocarcinogenic and promoting effects of asbestos.

IV. CONCLUSION

A host factor — genetic background, and a lifestyle factor — cigarette smoking, interact with occupational exposures in the production of human cancer. The mechanisms of these interactions have been elucidated by experimental studies.

For one kind of occupationally related bladder cancer, the inducing carcinogen is a chemical product, and genetics determine the characteristics of metabolism of the carcinogen. For a type of occupationally related lung cancer, the inducing carcinogen results from cigarette smoking and the enhancing factor is the occupational exposure to asbestos. Occupational exposures, therefore, can either contribute the causative agents whose action can be modified by other factors, or the causative agent can be a host-acquired element whose action is modified by occupational exposures (Table 3).

Lifestyle factors may impact directly on the occurrence of as much as 60% of human cancers in the U.S.[9] These factors include a diversity of elements (Table 4). Thus, the role of lifestyle factors in affecting risk from occupational exposures in particular requires serious attention.

Table 3
INTERACTION OF OCCUPATIONAL AGENTS WITH OTHER FACTORS

Genotoxic carcinogen	Enhancing factor	Example
Industrial product	Host	Bladder cancer in dye workers
Lifestyle or environmental agent	Industrial product	Lung cancer in asbestos workers who are cigarette smokers

Table 4
LIFESTYLE FACTORS AFFECTING CANCER RISKS

Diet
Voluntary practices
  Tobacco use
  Alcohol use
  Sunlight exposure
Customs
  Marriage and pregnancy
  Circumcision
Attitudes and behavior
  Perceptions of health maintenance
  Sexual hygiene
  Sexual behavior

ACKNOWLEDGMENT

I wish to thank Ms. J. Benvin for valuable help in the preparation of this manuscript.

REFERENCES

1. **Hueper, W. C.,** *Occupational Tumors and Allied Diseases,* Charles C Thomas, Springfield, Ill., 1942.
2. **Williams, M. H. C.,** Occupational tumors of the bladder, in *Cancer,* Vol. 3, Raven, R. W., Ed., Butterworths, London, 1958, 337.
3. **Holmberg, B.,** Setting of exposure standards, in *Advances in Medical Oncology, Research and Education, Epidemiology,* Vol. 3, Pergamon Press, New York, 1979, 115.
4. **Cole, P. and Goldman, M. B.,** Occupation, in *Persons at High Risk of Cancer: An Approach to Cancer Etiology and Control,* Fraumeni, J. F., Jr., Ed., Academic Press, New York, 1975, chap. 11.
5. **Williams, G. M., Reiss, B., and Weisburger, J. H.,** A comparison of the animal and human carcinogenicity of several environmental, occupational and therapeutic chemicals, in *Handbook of Experimental Pharmacology,* Andrews, L., Lorentzen, R., and Flamm, G., Eds., Springer Verlag, Berlin, in press.
6. **Hiatt, H. H., Watson, J. D., and Winsten, J. A., Eds.,** *Origins of Human Cancer, Human Risk Assessment,* Book C, Cold Spring Harbor Laboratory, Cold Spring Harbor, N.Y., 1977.
7. **Fraumeni, J. F., Jr., Ed.,** *Persons at High Risk of Cancer: An Approach to Cancer Etiology and Control,* Academic Press, New York, 1975.
8. **Bartsch, H. and Armstrong, B., Eds.,** *Host Factors in Human Carcinogenesis,* IARC Sci. Publ. No. 39, International Agency for Research on Cancer, Lyon, 1982.
9. **Weisburger, J. H. and Williams, G. M.,** Chemical carcinogenesis, in *Cancer Medicine,* 2nd ed., Holland, J. F. and Frei, E., III, Eds., Lea & Febiger, Philadelphia, 1982, 42.
10. International Agency for Research on Cancer, *IARC Monographs on the Evaluation of the Carcinogenic Risk of Chemicals to Humans,* IARC Monogr. Suppl. 4, IARC, Lyon, 1982.

11. **Weisburger, E. K.,** Metabolic studies in vivo with arylamines, in *Carcinogenic and Mutagenic N-Substituted Aryl Compounds,* NCI Monogr. 58, National Cancer Institute, Bethesda, 1981, 95.
12. **Weisburger, J. H. and Williams, G. M.,** Metabolism of chemical carcinogens, in *Cancer: A Comprehensive Treatise,* 2nd ed., Becker, F. F., Ed., Plenum Press, New York, 1982, 241.
13. **Weber, W. W. and Glowinski, I. B.,** Acetylation, in *Enzymatic Basis of Detoxification,* Vol. 2, Jakoby, W. B., Ed., Academic Press, New York, 1980, 169.
14. **McQueen, C. A., Maslansky, C. J., Glowinski, I. B., Crescenzi, S. B., Weber, W. W., and Williams, G. M.,** Relationship between the genetically determined acetylator phenotype and DNA damage by hydralazine and 2-aminofluorene in cultured rabbit hepatocytes, *Proc. Natl. Acad. Sci. U.S.A.,* 79, 1269, 1982.
15. **McQueen, C. A., Maslansky, C. J., and Williams, G. M.,** The role of acetylation polymorphism in determining susceptibility of cultured rabbit hepatocytes to DNA damage by aromatic amines, *Cancer Res.,* 43, 3120, 1983.
16. **Lower, G. M., Nilsson, T., Nelson, C. E., Wolf, H., Gamsky, T. E., and Bryan, G. T.,** N-Acetyltransferase phenotype and risk in urinary bladder cancer: approaches in molecular epidemiology. Preliminary results in Sweden and Denmark, *Environ. Health Perspect.,* 29, 71, 1979.
17. **Cartwright, R. A., Rogers, H. J., Barham-Hall, D., Glashan, R. W., Ahmad, R. A., Higgins, E., and Kahn, M. A.,** Role of N-acetyltransferase phenotypes in bladder carcinogenesis: a pharmacogenetic epidemiological approach to bladder cancer, *Lancet,* 2, 842, 1982.
18. **Williams, G. M., Weisburger, J. H., and Wynder, E. L.,** Lifestyle and cancer etiology, in *Carcinogens and Mutagens in the Environment,* Vol. 1, Stich, H. F., Ed., CRC Press, Boca Raton, Fla., 1982, chap. 5.
19. **Wynder, E. L., Bross, I. J., and Feldman, R. A.,** Study of etiological factors in cancer of the mouth, *Cancer,* 10, 1300, 1957.
20. **Rothman, K. J.,** Alcohol, in *Persons at High Risk of Cancer,* Fraumeni, J. F., Jr., Ed., Academic Press, New York, 1975, chap. 9.
21. **Selikoff, I. J., Hammond, E. C., and Churg, J.,** Asbestos exposure, smoking and neoplasia, *JAMA,* 204, 106, 1968.
22. **Saracci, R.,** Asbestos and lung cancer: an analysis of the epidemiological evidence on the asbestos-smoking interaction, *Int. J. Cancer,* 20, 323, 1977.
23. **Reiss, B., Solomon, S., Tong, C., Levenstein, M., Rosenberg, S. H., and Williams, G. M.,** Absence of mutagenic activity of three forms of asbestos in liver epithelial cells, *Environ. Res.,* 27, 389, 1982.
24. **Reiss, B., Tong, C., Telang, S., and Williams, G. M.,** Enhancement of benzo(a)pyrene mutagenicity by crysotile asbestos in rat liver epithelium cells, *Environ. Res.,* 31, 100, 1983.
25. **DiPaolo, J. A., DeMarinis, A. J., and Doniger, J.,** Asbestos and benzo(a)pyrene synergism in the transformation of Syrian hamster embryo cells, *J. Environ. Pathol. Toxicol.,* 5, 535, 1982.
26. **Topping, D. C. and Nettesheim, P.,** Two-stage carcinogenesis studies with asbestos in Fischer 344 rats, *J. Natl. Cancer Inst.,* 65, 627, 1980.
27. **Reiss, B., Solomon, S., Weisburger, J. H., and Williams, G. M.,** Comparative toxicities of different forms of asbestos in a cell culture assay, *Environ. Res.,* 22, 109, 1980.
28. **Wagner, J. C., Ed.,** *Biological Effect of Mineral Fibers,* Vols. 1 and 2, International Agency for Research on Cancer, Lyon, 1980.

Chapter 4

ASBESTOS AND NEOPLASTIC DISEASE*

Andrew Churg

TABLE OF CONTENTS

* Supported by Grant MT6907 from the Medical Research Council of Canada and a grant from the National Cancer Institute of Canada.

I. INTRODUCTION

Asbestos induces a wide variety of diseases, a large portion of which are neoplastic. At present asbestos is probably the leading cause of industrial-related malignancies, in large measure because of its widespread use in shipyards in World War II, during which time some 4 million persons were exposed in the U.S. alone.[1] The magnitude of the problem can also be appreciated from the fact that some 16,000 lawsuits, most claiming damages for an asbestos-induced tumor, are currently pending against the Manville Corporation,[2] and additional actions are pending against numerous other companies. In some series of highly exposed workers, 40 to 50% die from a malignancy.[3] This paper will very briefly consider the clinical, pathological, and epidemiological features of the various neoplasms which are associated with asbestos exposure; longer reviews can be found in References 4 to 6.

II. NATURE AND USE OF ASBESTOS MINERALS

Asbestos is a general term for a variety of silicate minerals which are naturally fibrous, and which have the additional properties of high heat resistance, high tensile strength, and reasonable imperviousness to chemical attack. The major types of asbestos are shown in Table 1.

Chrysotile accounts for more than 90% of the asbestos used worldwide, and most of the product comes from Quebec. The amphiboles are a large group of minerals, only a few of which are fibrous. Of these, only amosite and crocidolite are commercially useful. Tremolite, anthophyllite, and actinolite do not occur in concentrations or sizes of commercial utility, and are most often encountered as contaminants of other minerals, for example, talc or chrysotile ore. However, it should be remembered that the biologic properties of all the asbestos minerals are similar, regardless of their industrial uses.

Asbestos has a tremendous number of applications, largely in the production of construction materials (especially asbestos cement), insulating materials, friction products such as brake linings, and many common products such as floor and ceiling tiles. It has been estimated that there are more than 3000 uses for asbestos.[7] Asbestos minerals are also components of outcrop rock in many localities; the combination of extensive industrial use and natural rock weathering has resulted in asbestos contamination of the air and water supplies of many urban areas.[8] Data from our laboratory has shown that asbestos fibers can be recovered from the lungs of everyone in the population, whether or not they have a history of occupational asbestos exposure.[9] Although there are some reports suggesting that persons drinking asbestos-containing water have a higher incidence of malignancies,[10] the clinical significance of this widespread contamination is unknown.

III. ASBESTOS-INDUCED DISEASE

A. Asbestosis

As shown in Table 2, asbestos induces a variety of disease, generally of the lung and pleura. Asbestosis, or interstitial fibrosis of the lungs, was described at the beginning of the century. The disease is characterized[11] by progressive pulmonary impairment, usually presenting as shortness of breath on exercise. Radiographically such patients show a pattern of basilar interstitial infiltrates, and a restrictive pattern of disease on pulmonary function testing. Pathologically there is diffuse interstitial fibrosis accompanied by asbestos bodies, fibers of asbestos to which iron protein coat is added in the lung.[12] Despite strict dust control measures in plants processing or using asbestos, asbestosis is still not a rare disease, as can be seen in a recent report from Ontario where 39% of production workers in an asbestos cement factory were found to be suffering from this condition.[13] However, deaths from asbestosis

Table 1
MAJOR TYPES OF ASBESTOS MINERALS

Chrysotile (white asbestos)
Amphibole asbestos
  Crocidolite (blue asbestos)
  Amosite
  Tremolite
  Anthophyllite
  Actinolite

Table 2
DISEASES KNOWN OR THOUGHT TO BE ASSOCIATED WITH ASBESTOS EXPOSURE

Asbestosis (interstitial fibrosis of the lungs)
Fibrosis of small airways
Pleural plaques
Malignant mesothelioma of the pleura and peritoneum
Carcinoma of the lung
Carcinoma of the gastrointestinal tract
Carcinoma of the larynx
? Lymphomas

are now considerably outnumbered by deaths from asbestos-induced malignancies; in a recent study of asbestos insulation workers only 7% died of asbestosis, while 8% died of mesotheliomas, and 21% died of carcinoma of the lung.[3]

The importance of the fibrotic changes in asbestosis in the genesis of lung cancers are unclear. Some authors regard asbestos-associated bronchogenic carcinomas as a form of scar carcinoma, and diffuse interstitial fibrosis of any cause is known to be associated with an increased incidence of carcinoma,[14] but whether the incidence of cancer in lungs with asbestosis exceeds that in lungs with idiopathic interstitial fibrosis is unknown. However, it is possible that many such cancers in fact evolve in the lungs with no significant scarring.

B. Pleural Plaques

Pleural plaques are collagenous flat or knobbed structures found on the parietal and diaphragmatic pleurae; histologically they consist of bundles of virtually acellular collagen.[4] Plaques are most commonly detected radiographically, and the presence of calcified diaphragmatic plaques is considered pathognomonic of asbestos exposure. Plaques do not cause functional disease, and, although it has been suggested that patients with plaques are more prone to develop carcinoma of the lung,[15] the plaques themselves certainly play no role in this process, and the observation is itself disputed.[16] For the most part pleural plaques are merely markers of asbestos exposure. Asbestos also induces pleural effusions. These tend to be recurrent, and 80% are probably asymptomatic; effusions are seen relatively early after the onset of exposure, while the development of plaques requires longer time intervals.[17]

C. Mesothelioma

Asbestos exposure is associated with the development of malignant mesotheliomas of the pleura and peritoneum. Clinically, the pleural tumors typically present with chest pain and radiographic evidence of a pleural effusion, while the abdominal tumors may present with ascites, pain, or evidence of bowel obstruction. The diagnosis requires pathological ex-

Table 3
PERCENTAGE OF DEATHS DUE TO MESOTHELIOMA IN VARIOUS SERIES

Report	% deaths due to mesothelioma
Canadian chrysotile miners[20]	0.2
Australian crocidolite miners[21]	3.3
Amosite factory workers[22]	4.6
U.S.-Canadian insulation workers[3]	8
Nottingham gas mask assemblers[23]	10
Canadian gas mask assemblers[24]	16

amination of tissue and can be quite difficult histologically; in some instances autopsy is required to exclude another primary tumor.[4]

From an epidemiological and etiological point of view, mesothelioma shows the most clear cut and easiest to interpret association with asbestos exposure of any of the malignancies discussed in this paper. This is true because, with one disputed exception mentioned below, there are no epidemiologically defined causes of mesothelioma in humans other than asbestos; by contrast, there is an overwhelming association of lung and laryngeal carcinomas with smoking in all populations, and it is clear that there are numerous other as yet not clearly defined causes of gastrointestinal cancers as well.

There are numerous epidemiological studies showing the association of asbestos exposure and mesothelioma (see References 5, 6, and 18 for reviews). In the general population of North America, the incidence of malignant mesothelioma is approximately 1 to 2 cases per million persons per year.[19] In asbestos workers, mesothelioma is a fairly common tumor. As shown in Table 3, mesothelioma accounts for 10 to 16% of deaths in some series.[3,20-24] There is a dose-response relationship between the amount of exposure and the development of mesothelioma, however, it is clear that tumors appear at doses which are much lower than those required to produce asbestosis. Indeed, the presence of tumors in household contacts of asbestos workers has raised fears that very small amounts of asbestos might be carcinogenic.[25] The latent period for the development of mesothelioma is on average 30 to 40 years, and Peto et al.[26] have shown that tumor incidence is a function of time to the third or fourth power from the date of initial exposure.

Two important and unanswered questions with regard to mesothelioma are the importance of fiber type and fiber size. Chrysotile, at least in the form of mine dust, appears to be less dangerous than the amphiboles. For example, McDonald et al.[20] found a prevalence rate of 0.05% in Canadian chrysotile miners, compared to about 2% in South African crocidolite miners.[27] Crocidolite, in particular, is strongly carcinogenic in this regard, and amosite has been considered less so. This conclusion may, however, reflect different patterns of importation and use of the different asbestos fibers in different countries. Crocidolite was widely used in the U.K., whereas amosite was more extensively imported into North America. Actual mineralogic evaluation of lung tissue from mesothelioma cases in our own laboratory (unpublished data) clearly indicates a predominance of amosite in North America, and there is a high incidence of mesothelioma in amosite factory workers.[22] Anthophyllite has not been reported to produce tumors in humans.

Peto et al.[26] have suggested that fiber type also influences the distribution of pleural vs. peritoneal tumors, and that the latter are particularly associated with amphibole exposure. However, the data on this point are conflicting and fragmentary, and the effects might be reflections of dose rather than fiber type. Our own experience with tumors derived largely from shipyard workers is that pleural primaries outnumber peritoneal by about 10 to 1.

These data are in conflict with the results of experimental studies which produce roughly

Table 4
EFFECTS OF ASBESTOS AND CIGARETTE SMOKE IN INDUCING LUNG CANCER[37]

Group	Rel. risk of dying of lung cancer
Nonsmokers, no asbestos exposure	1.00
Nonsmokers, asbestos exposure	5.17
Smokers, no asbestos exposure	10.85
Smokers, asbestos exposure	53.24

equal incidences of mesothelioma with all types of asbestos when it is injected into the pleural cavity; indeed, in this system, any insoluble fiber of approximately asbestos size will induce tumors, and many of the nonasbestos forms such as fiberglass or potassium titanate are just as effective as asbestos.[28,29] It has been suggested from these studies that the most carcinogenic fibers are those with lengths greater than 8 μm and widths less than 0.25 μm, but this hypothesis is difficult to reconcile with measurements of dust sizes involved in human exposures, and is not supported by our own analysis of dust sizes found in human cases (unpublished). This question is discussed at length by Harington.[30]

The hypothesis that any type of very fine mineral fiber can induce mesotheliomas has received some support from the finding of an area in Turkey in which mesothelioma is endemic, and in which the soil and local building materials contain large amounts of the zeolite fiber, erionite.[31] This fiber is similar to asbestos in size, and produces asbestos-like ferruginous bodies in humans.[32] However, one recent study claims that there is also asbestos naturally present in the same region.[33]

D. Carcinoma of the Lung

Numerically, lung cancer is the major current health hazard associated with asbestos exposure. Doll and Peto[34] estimate that there were approximately 11,000 occupation-associated lung cancers in the U.S. in 1978, of which the vast majority were associated with asbestos exposure. This number represents about 10% of the total number of lung cancers for that year.

In the absence of asbestosis, there are no clinical or pathological features which distinguish asbestos-induced tumors from those arising purely on the basis of cigarette smoking.[4] The tumors typically arise 20 to 30 years after the onset of exposure. Statistically there is a slight predominance of lower lobe primaries, as opposed to upper lobe primaries in pure cigarette smokers, but the differences are small.[35] It has been claimed that asbestos-induced tumors are more commonly adenocarcinomas,[36] but this is disputed,[35] and all histologic types of lung cancer are common in the asbestos-exposed groups. It is important to bear these facts in mind when attempting to attribute a given case of lung cancer in a smoker to asbestos exposure.

Despite the major effects of cigarette smoke in producing most lung cancers, the superimposed effects of asbestos exposure have been repeatedly demonstrated (see Huuskonen,[18] Table 1 for a recent summary). It is clear that an excess risk exists for both smokers and nonsmokers, and that a more or less linear dose-response relationship applies. The exact interaction between cigarette smoke and asbestos is uncertain, and both additive and multiplicative models have been proposed. Table 4 shows data from Hammond et al.[37] which exactly fit a multiplicative model, and demonstrate an approximately 50-fold increase in lung cancer deaths over a nonexposed nonsmoking population. Table 5 shows data from the Quebec mining industry[20] which fall between the additive and multiplicative predictions.

The relative effects of different fibers in producing lung cancers in humans are unclear.

Table 5
RISK OF DYING OF LUNG CANCER FOR WORKERS IN THE QUEBEC CHRYSOTILE INDUSTRY[20]

	Asbestos exposure		
Smoking category	**Low**	**Moderate**	**High**
Nonsmokers	1	2.0	6.9
Moderate smokers	6.3	7.5	12.8
Heavy smokers	11.8	13.3	25.0

Some reports have suggested that chrysotile may be less carcinogenic than the amphiboles,[38] and there may be higher risks for those engaged in processing or insulating as opposed to mining. In a recent study of amosite factory workers, the risk for those who smoked was approximately 80 times that for a nonexposed nonsmoking control group.[22] This topic is briefly reviewed by McDonald.[6] However, it is certain that all types of asbestos including anthophyllite do produce tumors in humans,[39] in contradistinction to the situation for mesotheliomas.

E. Carcinoma of the Larynx

Considerably less data exist for carcinoma of the larynx, and both positive and negative studies have been published.[6,20,40-42] In the positive studies, relative risks have ranged from 1.07 for chrysotile miners to 14.5 for workers in the city of Liverpool. In all these reports, the strongest association for these tumors is cigarette smoking, and it is as yet unknown whether they appear in the absence of smoking.

F. Carcinoma of the Gastrointestinal Tract

GI carcinomas again represent tumors which are common in the general nonasbestos-exposed population, although the causative agent(s) is obscure. In some series of asbestos workers, relatively small increases have been seen, with ratios of observed to expected in the range of 1.5 to 3; in other series no excess has been observed.[3,6,18]

G. Lymphoma

In preliminary studies, it has been suggested that asbestos may also induce lymphoid and hematopoietic malignancies,[43] particularly of the oral cavity and GI tract. Another report has failed to demonstrate any such association.[44]

IV. SUMMARY

There is no doubt that asbestos exposure is associated with the development of a variety of malignancies, of which lung cancers are numerically the most important. A number of important questions about these associations remain. One of the most critical is the effects of low level exposure, since little is known about the dose-response curve in this area. Whether there is any "safe dose" is disputed. Hygiene standards for exposure have repeatedly been reduced with time, but it has been calculated that even a standard as low as 2 fibers per cubic centimeter of chrysotile would produce a 1% annual incidence of lung cancers after 50 years exposure.[45] Considerably more work is needed on the effects of fiber type and fiber size, since these parameters also go into the hygiene standards, and there appear to be considerable differences in effects for different types of asbestos usage.

REFERENCES

1. **Selikoff, I. J., Lilis, R., and Nicholson, W. J.,** Asbestos disease in United States shipyards, *Ann. N.Y. Acad. Sci.*, 330, 295, 1979.
2. **Feder, B. J.,** Manville submits bankruptcy filing to halt lawsuits — asbestos risks involved, *New York Times*, August 27, 1982.
3. **Selikoff, I. J., Hammond, E. C., and Seidman, H.,** Mortality experience of insulation workers in the United States and Canada, 1943—1976, *Ann. N.Y. Acad. Sci.*, 330, 91, 1979.
4. **Churg, A. and Golden, J.,** Current problems in asbestos-related disease, *Pathol. Ann.*, 17(Part 2), 33, 1982.
5. **Becklake, M. R.,** Asbestos-related diseases of the lung and other organs: their epidemiology and implications for clinical practice, *Am. Rev. Resp. Dis.*, 114, 187, 1976.
6. **McDonald, J. C.,** Asbestos-related disease: an epidemiological review, in *Biological Effects of Mineral Fibres*, Vol. 2, Wagner, J. C., Ed., IARC Sci. Publ. No. 30, International Agency for Research on Cancer, Lyon, 1980, 587.
7. **Hendry, N. W.,** The geology, occurrences and major uses of asbestos, *Ann. N.Y. Acad. Sci.*, 135, 12, 1965.
8. **Nicholson, W. J., Rohl, A. N., Weisman, I., and Selikoff, I. J.,** Environmental asbestos concentrations in the United States, in *Biological Effects of Mineral Fibres*, Vol. 2, Wagner, J. C., Ed., IARC Sci. Publ. No. 30, International Agency for Research on Cancer, Lyon, 1980, 823.
9. **Churg, A. and Warnock, M. L.,** Asbestos fibers in the general population, *Am. Rev. Resp. Dis.*, 122, 669, 1980.
10. **Kanarek, M. S., Conforti, P. M., Jackson, L. A., Cooper, R. C., and Murchio, J. C.,** Asbestos in drinking water and cancer incidence in the San Francisco Bay Area, *Am. J. Epidemiol.*, 112, 54, 1980.
11. **Medical Advisory Panel to the Asbestos International Association,** Criteria for the diagnosis of asbestosis and considerations in the attribution of lung cancer and mesothelioma to asbestos exposure, *Int. Arch. Occup. Environ. Health*, 49, 357, 1982.
12. **Churg, A.,** Fiber counting and analysis in the diagnosis of asbestos-related disease, *Human Pathol.*, 14, 381, 1982.
13. **Finkelstein, M. M.,** Asbestosis in long-term employees of an Ontario asbestos-cement factory, *Am. Rev. Resp. Dis.*, 125, 496, 1982.
14. **Turner-Warwick, M., Lebowitz, M., Burrows, B., and Johnson, A.,** Cryptogenic fibrosing alveolitis and lung cancer, *Thorax*, 35, 496, 1980.
15. **Edge, J. R.,** Incidence of bronchial carcinoma in shipyard workers with pleural plaques, *Ann. N.Y. Acad. Sci.*, 330, 289, 1979.
16. **Kiviluoto, R., Meurman, L. O., and Hakama, M.,** Pleural plaques and neoplasia in Finland, *Ann. N.Y. Acad. Sci.*, 330, 31, 1979.
17. **Epler, G. R., McLoud, T. C., and Gaensler, E. A.,** Prevalence and incidence of benign asbestos pleural effusion in a working population, *JAMA*, 247, 617, 1982.
18. **Huuskonen, M. S.,** Asbestos and cancer, *Eur. J. Resp. Dis.*, 63(Suppl. 123), 145, 1982.
19. **McDonald, A. C. and McDonald, J. C.,** Malignant mesothelioma in North America, *Cancer*, 46, 1650, 1980.
20. **McDonald, J. C., Liddell, F. D. K., Gibbs, G. W., Eyssen, G. E., and McDonald, A. D.,** Dust exposure and mortality in chrysotile mining, 1910—1975, *Br. J. Ind. Med.*, 37, 11, 1980.
21. **Hobbs, M. S. T., Woodward, S. D., Murphy, B., Musk, A. W., and Elder, J. E.,** The incidence of pneumoconiosis, mesothelioma and other respiratory cancer in men engaged in mining and milling crocidolite in Western Australia, in *Biological Effects of Mineral Fibres*, Vol. 2, Wagner, J. C., Ed., IARC Sci. Publ. No. 30, International Agency for Research on Cancer, Lyon, 1980, 615.
22. **Selikoff, I. J., Seidman, H., and Hammond, E. C.,** Mortality effects of cigarette smoking among amosite asbestos factory workers, *J. Natl. Cancer Inst.*, 65, 507, 1980.
23. **Jones, J. S. P., Smith, P. G., Pooley, F. D., Berry, G., Sawle, G. W., Madeley, R. J., Wignall, B. K., and Aggarwal, A.,** The consequences of exposure to asbestos dust in a wartime gas-mask factory, in *Biological Effects of Mineral Fibres*, Vol. 2, Wagner, J. C., Ed., IARC Sci. Publ. No. 30, International Agency for Research on Cancer, Lyon, 1980, 637.
24. **McDonald, A. D. and McDonald, J. C.,** Mesothelioma after crocidolite exposure during gas mask manufacture, *Environ. Res.*, 17, 340, 1978.
25. **Anderson, H. A., Lilis, R., Daum, S. M., Fischbein, A. S., and Selikoff, I. J.,** Household-contact asbestos neoplastic risk, *Ann. N.Y. Acad. Sci.*, 271, 311, 1976.
26. **Peto, J., Seidman, H., and Selikoff, I. J.,** Mesothelioma mortality in asbestos workers: implications for models of carcinogenesis and risk assessment, *Br. J. Cancer*, 45, 124, 1982.

27. **Talent, J. M., Harrison, W. O., Solomon, A., and Webster, I.,** A survey of black mineworkers of the Cape crocidolite mines, in *Biological Effects of Mineral Fibres,* Vol. 2, Wagner, J. C., Ed., IARC Sci. Publ. No. 30, International Agency for Research on Cancer, Lyon, 1980, 723.
28. **Wagner, J. C., Berry, G., and Timbrell, V.,** Mesotheliomata in rats after innoculation with asbestos and other materials, *Br. J. Cancer,* 28, 173, 1973.
29. **Stanton, M. F., Layard, M., Tegeris, A., Miller, E., May, M., Morgan, E., and Smith, A.,** Relation of particle dimension to carcinogenicity in amphibole asbestoses and other fibrous minerals, *J. Natl. Cancer Inst.,* 67, 965, 1981.
30. **Harington, J. S.,** Fiber carcinogenesis: epidemiologic observations and the Stanton hypothesis, *J. Natl. Cancer Inst.,* 67, 977, 1981.
31. **Baris, Y. I., Artvinli, M., and Sahin, A. A.,** Environmental mesothelioma in Turkey, *Ann. N.Y. Acad. Sci.,* 330, 423, 1979.
32. **Sebastien, P., Gaudichet, A., Bignon, J., and Baris, Y. I.,** Zeolite bodies in human lungs from Turkey, *Lab. Invest.,* 44, 420, 1981.
33. **Rohl, A. N., Langer, A. M., Moncure, G., Selikoff, I. J., and Fischbein, A. S.,** Endemic pleural disease associated with exposure to mixed fibrous dust in Turkey, *Science,* 216, 518, 1982.
34. **Doll, R. and Peto, R.,** The causes of cancer: quantitative estimates of avoidable risks of cancer in the United States today, *J. Natl. Cancer Inst.,*66, 1192, 1981.
35. **Kannerstein, M. and Churg, J.,** Pathology of carcinoma of the lung associated with asbestos exposure, *Cancer,* 30, 14, 1972.
36. **Whitwell, F., Newhouse, M. L., and Bennett, D. R.,** A study of the histological cell types of lung cancer in workers suffering from asbestosis in the United Kingdom, *Br. J. Ind. Med.,* 31, 298, 1974.
37. **Hammond, E. C., Selikoff, I. J., and Seidman, H.,** Asbestos exposure, smoking, and death rates, *Ann. N.Y. Acad. Sci.,* 330, 473, 1979.
38. **Hughes, J. and Weill, H.,** Lung cancer risk associated with manufacture of asbestos-cement products, in *Biological Effects of Mineral Fibres,* Vol. 2, Wagner, J. C., Ed., IARC Sci. Publ. No. 30, International Agency for Research on Cancer, Lyon, 1980, 627.
39. **Meurman, L., Kiviluoto, R., and Hakama, M.,** The mortality and morbidity among the working population of anthophyllite asbestos mines in Finland, *Br. J. Ind. Med.,* 31, 105, 1974.
40. **Shettigara, P. T. and Morgan, R. W.,** Asbestos, smoking, and laryngeal carcinoma, *Arch. Environ. Health,* 30, 517, 1975.
41. **Newhouse, M. L., Gregory, M. M., and Shannon, H.,** Etiology of carcinoma of the larynx, in *Biological Effects of Mineral Fibres,* Vol. 2, Wagner, J. C., Ed., IARC Sci. Publ. No. 30, International Agency for Research on Cancer, Lyon, 1980, 687.
42. **Stell, P. M. and McGill, T.,** Asbestos and laryngeal cancer, *Lancet,* 2, 416, 1973.
43. **Nichols, P. W., Ross, R., Dworksky, R., Wright, W., Koss, M., and Lukes, R. J.,** Large cell lymphomas in patients with occupational asbestos exposure, *Lab. Invest.,* 48, 63A, 1983.
44. **Olsson, H. and Brandt, I.,** Asbestos exposure and non-Hodgkin's lymphoma, *Lancet,* 1, 588, 1983.
45. **Peto, J.,** The hygiene standard for chrysotile asbestos, *Lancet,* 1, 484, 1978.

Chapter 5

RADIATION AS AN OCCUPATIONAL HAZARD

Gordon C. Butler

Since this paper* deals with carcinogens in the workplace I will restrict my remarks to the protection of workers, which is different from and in some ways easier to deal with than the protection of populations. The major topic with which I will deal in this chapter is the limitation of harm, or exposure or dose. Dose limitation for workers has two purposes: (1) to remove the risk of nonstochastic effects, e.g., skin erythema and opacity of the lens of the eye; (2) to reduce the incidence of stochastic effects, e.g., cancer and genetic effects, to acceptable levels and since there is no single view of what is acceptable it remains for a national authority, such as the Atomic Energy Control Board in Canada, to decide what levels are to be accepted.

The most important criterion of acceptability selected by the ICRP is that the average level of risks of injury and death to which atomic energy workers are subjected should be no greater than those for workers in "safe" industries,[2] e.g., manufacturing, which has an annual death rate from occupational hazards $= 1 \times 10^{-4}$.

This brings us to the estimation of risks from exposure to ionizing radiations. These have been provided since 1965 by the United Nations Scientific Committee on the Effects of Atomic Radiation (UNSCEAR) in its reports to the General Assembly[3] and, more recently, by the U.S. National Academy of Sciences through its Committee on the Biological Effects of Ionizing Radiation,[4] and by the ICRP itself.[2,5] These estimates are not greatly different.

For the most important radiosensitive tissues ICRP[1] adopted the following risk estimates, for purposes of radioprotection:

Gonads. The risk of serious hereditary ill health within the first two generations following the irradiation of either parent is taken to be about 10^{-4}/rem and the additional damage to later generations to be about equal to this. Since not all workers are of child-bearing age the genetically significant dose to a population will be less than the collective dose and is assumed by ICRP to be 40%. Therefore, the risk factor for gonads is taken to be 40×10^{-4}/Sv.

Breast. The female breast, during the reproductive period, is one of the most radiosensitive tissues for radiogenic cancer. The risk estimate will be both age- and sex-dependent. For radioprotection the risk factor is taken to be 25×10^{-4}/Sv.

Red bone marrow. Irradiation of red bone marrow can produce leukemia, for which the risk factor is taken to be 20×10^{-4}/Sv.

Lungs. External radiation can cause lung cancer in humans, and for this the risk factor is taken to be 20×10^{-4}/Sv.

Bone surfaces. Bone cancers from irradiation develop from cells on bone surfaces and this is the tissue specified for dosimetry. The risk factor is taken to be 5×10^{-4}/Sv.

* One thing that should be made clear at the beginning of this paper is that many of my statements will reflect, and in some cases will be identical with those of the International Commission on Radiological Protection (ICRP). It is exceptional in the subject of protection against carcinogens in the workplace to find an international authority that gives leadership in the subject and which enjoys such a wide acceptance. The ICRP has enunciated three basic principles for radiation protection:[1] (1) no practice shall be adopted unless its introduction produces a positive net benefit; (2) all exposures shall be kept as low as reasonably achievable, economic and social factors being taken into account; (3) the dose equivalent to individuals shall not exceed the limits recommended for the appropriate circumstances by the Commission.

Table 1
RISK COEFFICIENTS, f_T, FOR STOCHASTIC EFFECTS AND DERIVED WEIGHTING FACTORS RECOMMENDED BY THE ICRP

Tissue T	f_T (10^{-4}/Sv)	w_T ($= f_T/f_{WB}$)
Gonads[a]	40	0.25
Breast[a]	25	0.15
Red bone marrow	20	0.12
Lungs	20	0.12
Bone surfaces	5	0.03
Thyroid	5	0.03
Remainder	50	0.30
Uniform whole-body irradiation	165	1.00

[a] Averaged over both sexes and all ages.

Table 2
RISK COEFFICIENTS FOR CLASSES OF EFFECTS

Effect	Risk factor (10^{-4}/Sv)
Fatal cancers	100
Nonfatal tumors	
Breast	15
Thyroid	100
Skin	100
Total	215
Genetic defects	
First two generations	40
Total	80

Thyroid. The sensitivity of the thyroid gland to the induction of cancer is somewhat greater than the red bone marrow but only a fraction of the cancers prove to be fatal. The risk factor is therefore taken to be one fourth that for red bone marrow, namely 5×10^{-4}/Sv.
Remainder. There are other radiogenic cancers resulting from whole body irradiation but numerical estimates for these other tissues, such as the GI tract and liver, are not available. It is assumed that it does not total more than 50×10^{-4}/Sv.

These risk factors and the weighting factors derived from them are summarized in Table 1.

One can summarize these data in another way, as shown in Table 2. These estimates of Table 2 will be somewhat different for workers and members of the general public.

All radiation exposures are unlikely to be of the whole body uniform type; they may consist of partial body exposures from both external and internal sources. Providing that each can be estimated they can be listed and multiplied by the appropriate weighting factor to give a list of weighted doses. The sum of these gives the Effective Dose Equivalent (EDE), i.e.,

$$\sum H_T \cdot W_T = EDE \tag{1}$$

The ICRP recommendation[1] is that the EDE received in a year should not exceed 5 rems or 0.05 Sv.

It has been found from available statistics collected for groups of workers that a dose limit of 5 rems for a group of workers will result in an average dose to the group of one tenth the limit, or 0.5 rems, and this is what is used to calculate the risk to the group. From Table 2 it can be seen that this will result in an annual risk of fatal cancer of 0.5×10^{-4}/rem and a risk of genetic defects of 0.5×10^{-4}/rem, a total of 1×10^{-4}/rem, equivalent to the risk of death from occupational hazards in "safe" industries.

This annual dose limit on the EDE is the same as the previously recommended limit of 5 rem for whole body radiation. It has stood the test of time and, aside from a few disputes about details, it can be assumed that it provides an acceptable level of protection for the individual worker.

The EDE approach and the numerical limit are likely to find their way into revised dose limits, including those soon to be promulgated in Canada. It would be convenient if it could be used for the protection of all radiation workers, including miners.

There are, however, special problems with the radiological safety of miners, especially uranium miners in underground mines. These workers are exposed to the inhalation of radon and its daughters, thoron and its daughters, silica dust, and the exhaust fumes of diesel engines. In addition, in mines containing high grade ore the workers are exposed to γ-rays and β-particles external to the body. There is a sad history of more than 50 years of the occurrence of lung cancer in some of these miners who have received elevated exposures, usually measured in the concentration of radon daughters in the air. This is the accepted major cause of the cancer, but no one knows for certain the contribution from the other contaminants of mine air and the habit of cigarette smoking with an estimated practice in more than one half of the miners studied.

Let us concentrate on the problem of protecting the miners against the sources of ionizing radiation. It is desirable that they should be afforded the same degree of protection against the effects of radiation as other radiation workers, i.e., an EDE limit of 0.05 Sv/year. The difficulty is that there are too many uncertainties in calculating the EDE resulting from exposure to radon and its daughters. These uncertainties arise from the physical state of the daughter atoms, the respiratory physiology of the worker, and the cytology of the cells at risk. In spite of these difficulties estimates have been made which vary from 0.5 to 1.0 rad/WLM.[6] More recently, Johnson and Leach[7] give a value of 0.7 rad/WLM.*

In addition to this uncertainty there is the problem of choosing an appropriate value for the quality factor for α-particles. Values of 5 to 10 have been suggested by UNSCEAR[6] and 20 has been used by the ICRP.[1]

Finally there is the problem of the appropriate weighting factor which is derived from the risk estimates, of which more is to be said later.

In estimating the EDE for thoron and its daughters there are the problems mentioned above for radon + daughters plus the difficulties in calculating the doses to body tissues other than lungs, resulting from the absorption and diffusion of thoron in the body.

For these reasons the "dosimetric approach" cannot yet be taken with confidence and much attention has been devoted to the "epidemiological" approach. Some studies of the incidence of lung cancer in miners have yielded risk estimates that lie in the range 2 to 4.5 $\times 10^{-4}$/WLM (UNSCEAR[6]), 2.5 to 4.5 $\times 10^{-4}$/WLM (ICRP[8]), 1 to 6 $\times 10^{-4}$/WLM

* A unit of exposure used in the protection of miners, which results from breathing a concentration of 1 working level (WL) for a working month (170 hr). The WL corresponds to a concentration of 100 pCi/ℓ of ^{222}Rn in equilibrium with its daughters. 1 WL = 1.3×10^{-5} Mev/ℓ = 2.08×10^{-5} J/m^3 = 3.7 Bq/ℓ = 3700 Bq/m^3. For ^{220}Rn and its daughters in equilibrium 1 WL = 7.43 pCi/ℓ = 0.275 Bq/ℓ = 275 Bq/m^3 (see Reference 8).

(AECB[9]), and $> 1.2 \times 10^{-4}$/WLM (OML[10]). If one chose a mean number from these ranges it might be 2.5, although the NCRP[11] in a recent study seems to favor a somewhat lower number. Therefore in round numbers one could use the risk estimate 2×10^{-4}/WLM. It should be noted that this is the risk of contracting a fatal lung cancer. The present exposure limit in North America is 4 WLM/year. A recent study of Ontario miners[10] shows that almost none work more than 10 years underground, and that the average exposure in Ontario mines is about 2 WLM/year. Thus the risk to these miners is 40×10^{-4}.

Another way of assessing the risk estimate of 2×10^{-4} fatal lung cancers per WLM is to compare it with the risks of harm resulting from whole body irradiation shown in Table 2. This table shows that 1 rem (0.01 Sv) will result in a risk of fatal cancer of about 1×10^{-4} plus a risk of nonfatal tumor of 2×10^{-4} and of genetic defects of 1×10^{-4}. Exposure to radon and its daughters is not known to result in any harm other than fatal lung cancers. Thus, while exposure of miners to radon and its daughters in mine atmospheres appears to be twice as harmful in inducing fatal cancers it is less harmful *in toto*. A problem facing us at the moment is how to weight these nonfatal effects with respect to the fatal carcinogenic effects and there are differences of opinion on the point.

In its Publication 32, the ICRP[8] has proposed that the annual exposure limit for ^{222}Rn and its daughters should be 5 WLM, for ^{220}Rn it should be 14 WLM, and that

$$\left(\frac{H_E}{H_{E,lim}}\right)_{ext} + \sum \frac{I}{I_{lim}} \leqq 1 \qquad (2)$$

where H_E = the EDE of all external radiation; $H_{E,lim}$ = the EDE limit of 0.05 Sv; I = the exposure to each inhaled α-emitter, including radon + daughters, thoron + daughters, dusts containing long-lived α-emitting radionuclides; and I_{lim} = the annual limit on exposure to the above airborne α-emitters.

Let us look at what effect this recommendation might have on a typical Ontario mine where the exposures to γ-rays are about 1 rem/year, to ^{220}Rn about 2 WLM/year, and to long-lived α-emitters (airborne) about 12% of the annual limit on intake (ALI). This leaves about one half of the exposure limit (2.5 WLM) for radon daughters.[12]

It is impossible to evaluate the impact of risks from inhaling airborne α-emitters in mine atmospheres without considering the statistics of total accidents in this occupation. For example, in a group of Ontario miners studied over some years there was an excess of violent deaths which amounted to about five times the excess of lung cancer deaths. These statistics do not alter the fact that miners need the same degree of radiation protection as other workers, but it has some bearing on how the moneys available for all safety measures may be apportioned.

Finally, there is much interest in the risks from inhalation of radon daughters inside the home. For many reasons the doses and risk estimates for this group will be different from those for miners and the risk estimates for miners should not be used to estimate risks to populations.

REFERENCES

1. Recommendations of the International Commission on Radiological Protection, ICRP Publ. 26, *Ann. ICRP*, 1(3), 1977.
2. Problems involved in developing an index of harm, ICRP Publ. 27, *Ann. ICRP*, 1(4), 1977.
3. United Nations Scientific Committee on the Effects of Atomic Radiation, *Ionizing Radiation: Sources and Biological Effects*, 1982 Report to the General Assembly, United Nations, New York, 1982.
4. NAS Advisory Committee on the Biological Effects of Ionizing Radiations, *The Effects on Populations of Exposure to Low Levels of Ionizing Radiation: 1980*, (BEIR III), National Academy Press, Washington, D.C., 1980.
5. The evaluation of risks from radiation, ICRP Publ. 8, *Health Phys.*, 12, 239, 1966.
6. United Nations Scientific Committee on the Effects of Atomic Radiation, *Sources and Effects of Ionizing Radiation*, 1977 Report to the General Assembly, United Nations, New York, 1977, 398 (paragraph 222).
7. **Johnson, J. R. and Leach, V. A.**, An examination of the relationship between WLM exposure and dose, in *Radiation Hazards in Mining: Control, Measurement, and Medical Aspects*, Gomez, M., Ed., Society of Mining Engineers, New York, 1982, 390.
8. Limits for inhalation of radon daughters by workers, ICRP Publ. 32, *Ann. ICRP*, 6(1), 1981.
9. *Risk Estimates for Exposure to Alpha Emitters*, ACRP-2, AECB Publ. Info. 0090, Atomic Energy Control Board, Ottawa, 1982.
10. **Müller, J., Wheeler, W. C., Gentleman, J. F., Suranyi, G., and Kusiak, R. A.**, *Study of Mortality of Ontario Miners 1955—1977, Part I*, Report of a study supported jointly by the Ontario Ministry of Labour, the Ontario Workers' Compensation Board, and the Atomic Energy Control Board of Canada, Ontario Ministry of Labour, Toronto, 1983.
11. **Sinclair, W. K.**, Private communication, 1983.
12. **Myers, D. K.**, Private communication, 1983.

Part B
Tissues in Cancer Prevention

Chapter 6

IMPROVEMENTS FOR WORKER PROTECTION

John C. Marshall

TABLE OF CONTENTS

I. INTRODUCTION

It is now more than 200 years since the British surgeon, Sir Percival Pott observed that scrotal cancer occurred with unexpected frequency among chimney sweeps, and "... seems to derive from a lodgement of soot in the rugae of the scrotum."[1] Pott's observations marked a milestone in our understanding of the causes and prevention of cancer. Not only did he for the first time link work with the subsequent occurrence of cancer, but he recognized a more basic principle: that the human organism interacts biologically with its environment, and that that interaction can, in certain circumstances, produce disease.

Pott's work occurred in the historical context of 18th century England. Child labor was the order of the day. Young boys, most between the ages of four and seven, were apprenticed to master chimney sweeps in a manner that bordered on slavery. Their principal asset was their size, and their job was to physically enter the narrow brick chimneys to clean them. This was dangerous, brutal work, and the literature from the period abounds with stories of the deaths of children, suffocated when they became trapped in the chimneys, or badly burned when they were forced to enter the still-hot chutes so that their wealthy clients did not have to suffer too long the effects of an unheated room. Those who survived the ordeal were found to be prone, many years later, to cancers of the skin and scrotum, and Pott astutely deduced that these were a result of the effects of soot absorbed through their work clothes.

This particular chapter in the history of occupational carcinogenesis has become a paradigm for the problem in the 20th century. Cancer of the scrotum in chimney sweeps is of historical interest only. Changes in legislation with the abolition of child labor in the 19th century, and changes in technology with the demise of the coal fire as a method of heating have all but eliminated the problem. On the other hand, the same class of carcinogens that Pott identified has become the single most important preventable cause of cancer in our time. The complex aromatic hydrocarbons found in soot are widely encountered today in cigarette smoke, coke oven emissions, urban air pollution, as well as in a host of other processes, both industrial and nonindustrial, that involve the combustion of organic matter. The difference in cancer patterns reflects changes in social and personal mores, changes in industrial technology, and changes in social legislation.

II. BACKGROUND: THE ORIGINS OF OCCUPATIONAL CANCER AND THE DIMENSIONS OF THE PROBLEM

The question of what has been done, and more importantly, what still needs to be done to protect the worker from occupational carcinogens cannot be viewed abstracted from its social and political context. It is true that industrial cancer has its roots in industrialization, but at the same time, it is an important concept in understanding the nature of the problem today. With few exceptions, the process of evolution has created a remarkable harmony between life and the environment, and naturally occurring carcinogens are distinctly unusual.

The advent of the Industrial Revolution saw a significant change in both the nature of production and the organization of work. Materials from the environment were transformed on a scale never before seen. New substances were created, old substances assumed new forms, and a previously unimagined demand for new energy sources arose. The period is typified by Dickensian images of smoking chimneys, machinery, and crowded, dirty cities. Simultaneously the organization of work changed profoundly: the factory, the mine, and the mill replaced the farm as the central economic unit, and with this came the organization of the trade union. With this as well came an increasing awareness of the adverse effects of coal smoke and soot, mercury, arsenic, and metal fumes.

The 20th century has been the century of the petrochemical industry. From refining substances in the environment, we have moved to creating entirely new chemicals from the building blocks of fossil fuels. Over a million entirely synthetic chemicals have been created, and until recently, thousands more appeared each year. They have transformed our lives by providing us with plastics and synthetic fabrics, foods, potent pharmaceuticals, and powerful pesticides that have greatly aided agricultural production; however, these miracle substances have not been obtained without a price. They have transformed the environment in ways which are potentially disastrous, and they have taken their toll in disease. We are only beginning to see their darker side, exemplified by the vinyl chloride saga, for it was only in the early 1960s that petrochemical production began to mushroom.

The second half of the 20th century has also been the atomic age, and although the threat of global destruction has been foremost in our perception of the darker side of this technology, the carcinogenic effects of radiation are all too well known.

A consideration of what can be done to protect the worker must therefore situate itself in the context of modern industry: workers today are faced with thousands of potentially toxic chemicals, most unidentified, and most untested. When we look at the problem of identifying carcinogens in this vast chemical wasteland, the situation is further confounded. Occupational cancers become manifest only after a latency of several decades; for many substances, only small groups of workers are exposed, making valid epidemiological conclusions difficult. Add to this the fact that the majority of occupational cancers involve common sites, lung, skin, and bladder, and it is no surprise that the scientific state of the art is so primitive.

III. RESPONSES TO THE PROBLEM

Until quite recently, the problem of occupational cancer has been accorded very little consideration in social consciousness and in legislation. There is ample evidence to suggest that the workplace is second only to cigarette smoking as an identifiable and controllable cause of cancer mortality, yet in Canada there is no national policy on occupational cancer, and the workplace as a cause of cancer has been all but ignored in mass cancer education programs. It is a recurring theme that what little has been done has been largely a consequence of initiatives taken by workers and their unions, and that society at large, and the scientific community in particular, have been slow to accord the problem its proper due.

The prevention of occupational cancer involves four interrelated processes: (1) the identification of carcinogens; (2) education and the dissemination of this knowledge; (3) the development of new technologies to eliminate those problems identified; and (4) the legislative will to see changes introduced and enforced. These will be reviewed in turn.

A. Identification of Carcinogens

The science of epidemiology has been the cornerstone of our understanding of the carcinogenic effects of industrial agents. For the most part, the application of epidemiology to occupational carcinogenesis has been a rear guard action, initiated only when problems have been sufficiently dramatic that they have been apparent to even the casual observer. In other words, a strong suspicion that something was wrong has been the trigger for a more vigorous epidemiological investigation. Often, too, the impetus for scientific research has arisen not within the scientific community but elsewhere.

Selikoff's ongoing landmark study of asbestos insulation workers came about as a result of the moral and financial support of the insulators union in the U.S.[2] Much of our present knowledge about asbestos can be attributed to results from this study, or studies inspired by Selikoff's work. In Canada, the efforts of the United Steelworkers were the impetus for our knowledge about the relationship between nickel and sinus cancer in Sudbury, or between radon daughters and lung cancer in Elliot Lake. We learned that BCME was a potent lung carcinogen only as a result of the persistent efforts of workers at Rohm and Haas in the

U.S. Similarly, much of the ongoing research in the petrochemical industry has been initiated by the Oil, Chemical, and Atomic Workers and the United Rubberworkers.

This is not to deprecate the epidemiologist, but merely to illustrate that we have applied epidemiology by and large in reaction to problems that first made their appearance in the political and social arena. There are important exceptions: Creech's astute observations when confronted with two cases of angiosarcoma in Louisville, Kentucky led to the identification of vinyl chloride as a carcinogen.[3]

To apply epidemiology to the protection of the worker requires taking a more aggressive role in seeking out associations between work and cancer. To a limited extent, this has been done. The National Cancer Institute in the U.S. examined cancer rates geographically, and has documented marked variations in the rates of many common cancers from county to county or state to state, suggesting a key role for environmental factors. Some of these findings have ready explanations: the clustering of bladder cancers in the northeast reflects the distribution of the chemical industry there, and the elevated rates of lung cancer along the Gulf Coast have been attributed to the effects of asbestos in the shipbuilding industry. Mencke reviewed lung cancer rates by occupation in Los Angeles County (California) and found significantly elevated rates in such previously unsuspected occupational groups as dental lab technicians, decorators, taxi drivers, mechanics and photoengravers.[4] Geographic pathology, admittedly, does not allow for an analysis of the effects of such confounding variables as cigarette smoking, but it does provide an alert which should trigger further investigations. Milham's[5] exhaustive work in the northwestern U.S. will be reviewed later in Volume 5, and is another example of the kind of information that can be derived from actively seeking out problems.

One of the most exciting advances in recent years has been the adoption by workers of some of the principles of epidemiology. The United Auto Workers have developed a program of worker epidemiology and have produced a manual called "Tracking Down Workplace Killers: A Manual for Cancer Detectives on the Job".[6] Union health and safety representatives are trained to conduct proportional mortality ratio studies using seniority lists, death certificates, and mortality statistics from the general population. Their work has already begun to point to problems previously unrecognized in the auto industry. In Canada, the Ontario Federation of Labour has taught the basics of epidemiology to thousands of worker representatives using a simple card game called the "Cancer Game".[7] The principal value of such initiatives lies not in generating scientific data for publication, but in alerting workers to potential problems, and, more importantly, in demystifying what has previously been an arcane and impenetrable field of scientific endeavor.

The epidemiological approach to identifying carcinogens has come under attack as it occurs after the fact, once the damage is done. In the future, we need to move to a greater reliance on the pretesting of chemicals using animal systems, applying the same standards to industrial chemicals as are now applied to food and drugs. Some significant advances have been made in this area in recent years, and the utility of such an approach is amply demonstrated by the fact that of all human carcinogens now recognized, only arsenic is not clearly carcinogenic in animal systems. Interestingly, the impetus for such pretesting is coming not because of legislation, but because of fears in industry of litigation.

B. Education

The gulf between scientific knowledge and public awareness, as any educator knows, is very wide, but if there is a single key to protecting workers, it is education.

Unfortunately, ignorance abounds. Physicians are commonly seen as one of the key elements in health education, but it is not uncharitable to say that most doctors' knowledge in this field is lacking. Even in such well-documented areas as the effects of smoking, much of our knowledge derives not from medical curricula, but from the efforts of such agencies as the Cancer Society. When it comes to an awareness of occupational carcinogens, or the taking of an adequate occupational history, the average doctor is terribly deficient.

This ignorance is compounded by the awesome complexities of retrieving information from the scientific literature. Important information on the carcinogenic potential of such compounds as vinyl chloride and DBCP lay unheeded for years in scientific journals before human data brought it into focus.

However, significant advances have been made in the area of public education. Most notable in this regard have been the media, who despite occasional excesses and inaccuracies, have brought the problem to national and international awareness, and served as a catalyst for needed research and legislative changes. Contrary to fears that media publicity would lead to fear, hysteria, and political turmoil, the result of the past decade has been responsible legislation rather than civil unrest. Much of the pioneering work of OSHA under the Carter administration was facilitated by its prominence in the media. In Canada, the dramatic plight of workers appearing before the Ham Commission in Ontario proved to be the ultimate catalyst in bringing about comprehensive health and safety legislation.

Important educational programs have also appeared from within the trade union movement. Prominent among these is the program of the Ontario Federation of Labour which offers a three week, full-time course in the basics of health and safety for unionists.[7] Students become versed in such areas as toxicology, physiology, epidemiology, and engineering, as well as in more traditional areas such as collective bargaining and legislation. The program is conducted entirely using instructors drawn from the ranks of workers, and the level of sophistication of the average graduate is high. Much to the chagrin of unsuspecting employers and the occasional embarrassment of health and safety professionals, graduates have found themselves able to intelligently critique the scientific literature and to recommend practical improvements in the workplace. Worker education has assumed a growing role in community college programs as well. An informed, responsible workforce is the result.

Conspicuous by their absence have been the public education system and voluntary organizations such as the Cancer Society. There is an urgent need to develop a focus on the workplace as a cause of cancer to meet the demands of an increasingly knowledgeable public.

C. Technological Change

While the identification of occupational carcinogens lies in the scientific realm, the control of the problem lies in the sphere of technology. In the past, the control of worker exposure to industrial substances played a secondary role in the design of industrial processes. That situation is changing.

The relative ease with which the chemical industry, with a minimum of expense, was able to reduce vinyl chloride exposure from 500 ppm to 1 ppm is a striking example of the feasibility of technological solutions. In another area, many substitutes have been found for asbestos in situations where the mineral was previously considered indispensible. The development of ceramic disc brakes and new insulating materials are but two examples.

Technological change has resulted more from an economic imperative than from the pressures of legislation. A classic example of this is to be found in the steel industry. In North America, lung cancer rates in coke oven workers range from 10 to 20 times the national average.[8] Exposures to benzpyrene, the equivalent of smoking more than 1000 cigarettes a day have been found in industries where the physical facility often dates from the turn of the century. After World War II, in the process of rebuilding a devastated economy, Japan and Germany have developed some of the most advanced, safest coke oven technology to be found by the use of automation in Japan and improved coke oven design in Germany.

Fear of litigation has been a potent stimulus to the development of asbestos substitutes. In fact, the threat of litigation may well emerge in the 1980s as the dominant force leading to change in the workplace. In the face of a failing economy and record unemployment, governments and unions have been assuming a more conservative posture, cowed by the specter of the loss of jobs. At the same time, fear of litigation has stimulated changes not only in the asbestos industry, but in the plastics industry as well. Reports are beginning to emerge of an exhaustive pretesting program currently underway for a chemical hailed as a safe substitute for vinyl chloride.

D. Legislation

Comprehensive occupational health and safety legislation is a recent phenomenon. Although workers compensation legislation was introduced shortly after the turn of the century, it has only been in the 1970s that most Canadian provinces have introduced comprehensive health and safety legislation that can begin to provide workers with necessary legal protection. Robert Sass, former director of the Occupational Health and Safety Division in the province of Saskatchewan has identified three basic rights that health and safety legislation must provide: (1) the right to know, (2) the right to participate, and (3) the right to refuse unsafe or unhealthy work.[9]

These rights have been incorporated, to varying degrees, in most jurisdictions, but even today they are still lacking in the province of Nova Scotia.

In those provinces where legislation is in place, enforcement has been a problem, though the establishment of joint health and safety committees has provided a limited form of internal policing. A measure of the failings of implementation of legislation can be seen in Ontario's Workers' Compensation legislation. If we accept the conservative estimates of Peto and Doll[10] of 4% of cancer being attributable to occupational exposure, then only 1 of every 17 of these cases — fewer than 6% — receives compensation.

Legislation plays a role beyond providing legal protection for workers; it codifies what society feels to be acceptable standards of behavior and practice. It is the embodiment of our resolve to correct a problem. In this regard, many deficiencies exist for we have not yet stated as a society that we will accord the worker the right to a cancer-free workplace. To do so remains a priority for coming years, specifically, to develop a cancer policy covering the pretesting of chemicals introduced into industry, labeling and disclosure of the identities of substances already present, and the institution of performance standards to ensure that exposure to known carcinogens is kept to a minimum.

IV. CONCLUSION

This, then, has been a brief review of what has been accomplished in the area of worker protection, and what still needs to be done.

When one looks at the problem of workplace cancer one is struck by how much, yet how little has changed. Pott's chimney sweeps have passed into history and their soot warts have become historical vignettes. Yet, contemporary studies of chimney sweeps show that they now suffer elevated rates of lung and esophageal cancer, a reflection of changing technology.[11,12]

Much has been accomplished in the past decade, but those who have been at the forefront have labored in an atmosphere of mistrust and controversy. Occupational cancer is preventable cancer. The challenge facing us in the next decade is to move the problem to center stage, and to apply the same resources and principles that have been so widely used in our attack on lifestyle cancers to its ultimate eradication.

REFERENCES

1. **Pott, P.,** Chirurgical Observations Relative to the Cataract, Polypus of the Nose, the Cancer of the Scrotum, the Different Kinds of Ruptures, and the Mortification of the Toes and Feet, London, 1775.
2. **Selikoff, I. J., Hammond, E. C., and Seidman, H.,** Mortality experience of insulation workers in the United States and Canada, 1943—1976, *Ann. N.Y. Acad. Sci.*, 330, 91, 1979.
3. **Creech, J. L. and Johnson, M. N.,** Angiosarcoma of the liver in the manufacture of polyvinyl chloride, *J. Occup. Med.*, 16, 150, 1974.
4. **Mencke, H. R. and Henderson, B. E.,** Occupational differences in rates of lung cancer, *J. Occup. Med.*, 18(12), 797, 1976.
5. **Milham, S., Jr.,** Neoplasia in the wood and pulp industry, *Ann. N.Y. Acad. Sci.*, 271, 294, 1976.
6. United Auto Workers, *Tracking Down Workplace Killers: A Manual for Cancer Detectives on the Job,* UAW, Detroit, 1981.
7. Ontario Federation of Labour, *Occupational Health and Safety: A Training Manual,* Copp Clark Pitman, Toronto, 1982.
8. **Lloyd, J. W.,** Longterm mortality study of steelworkers: respiratory cancer in coke plant workers, *J. Occup. Med.*, 13, 53, 1971.
9. **Sass, R.,** Occupational health and safety right, unpublished speech, 1978.
10. **Doll, R. and Peto, R.,** The causes of cancer: quantitative estimates of avoidable risks of cancer in the United States today, *J. Natl. Cancer Inst.*, 66(6), 1191, 1981.
11. **Hansen, E. S.,** Mortality of Danish chimneysweeps 1970—1975, in Proc. Inst. Occup. Health Int. Symp. Prevention of Occupational Cancer, Helsinki, April 21—24, 1981, p. 104.
12. **Hogstedt, C., Andersson, K., Frenning, B., and Gustavsson, A.,** A cohort study on mortality among long-time employed Swedish chimneysweeps, *Scand. J. Work Environ. Health,* 8(Suppl. 1), 72, 1982.

Chapter 7

GOVERNMENT, EMPLOYERS, LABOR: INTERACTIONS THROUGH INFORMATION AND PARTICIPATION

Gordon Atherley and Robert Whiting

TABLE OF CONTENTS

I. INFORMATION

The basis of control of occupational cancer, whether by regulation or other means, has been trust that those few organizations and individuals with access to essential information would exercise their responsibilities diligently and equitably. Confidence in this relatively closed system has been undermined by the failure to prevent foreseeable occupational cancers, by problems in the quality control of scientific information, for example at the Industrial Biotest Laboratories, and by continuing controversies about policies for regulating carcinogens. Dissatisfactions have led to demands for more and better information, and to calls for more open decision-making processes involving government, employers, and workers.

New approaches are certainly required because information about occupational carcinogens is often uncertain, mistrusted, or controversial, and often unavailable to workers and employers. Moreover, only a few major carcinogens have been regulated, resulting in a belief that regulation cannot cover all risks of occupational carcinogenesis.

The key to effective new approaches to controlling carcinogens is confidence: in the relevance and quality of scientific data, in our ability to predict cancer risks, in the openness and balance of the regulatory process, and in the fairness of the burdens of responsibility on government, employers, and labor.

Confidence was shaken, for example, in 1976 by the report of the Ontario Royal Commission on the Health and Safety of Workers in Mines[1] which documented the occurrence of lung cancers in Ontario uranium miners, and the failure of government and industry in Ontario in the 1950s and 1960s to use available information and technology to prevent the cancers. In fact, much information about the cancers only became public during the inquiry. The report stressed that:

> "The responsibility-system seems to have been lacking in two significant ways. First, divided jurisdictions have made it unclear where the initiative necessary to deal with problems is to be taken. Second, the worker as an individual and workers collectively in labour unions or otherwise have been denied effective participation in tackling these problems; thus the essential principles of openness and natural justice have not received adequate expression."

The goals of *openness* and *participation* were fundamental to the establishment of the Canadian Centre for Occupational Health and Safety (CCOHS) by an Act of the Parliament of Canada in 1978. The Centre is an independent tripartite organization with a broad mandate to promote occupational health and safety, but without regulatory functions. Its policies and programs are determined by a Council of Governors representing federal, provincial and territorial governments, employers, and labor.

The Council decided that the Centre's first priority was the provision of an information and advisory service complemented by the development of computerized systems for rapid access to information. Through the service, information and advice is available to workers, managers, employers, governments, and others on virtually any topic in occupational health and safety. In establishing this priority, the Centre recognized that occupational health and safety could not progress unless all those involved — whether government, employers, or labor — have access to the best available information and advice. The spectacular growth in demand for the service shows the reality of the need.

Information and the flow of information about occupational cancer and carcinogens can be pictured at four interconnected levels:

(1) *Data* comprising research results, statistics, workplace exposure measurements, medical records, and occupational histories. Government, employers, and labor all have unique yet often uncoordinated sources of data which together could contribute more to the understanding and prevention of cancer.

(2) *Assessments and interpretations* of data, contained in reviews, reports, and textbooks.
(3) *Evaluations and recommendations* by organizations and individuals, often representing government, employers, or labor.
(4) *Control measures, regulations, and exposure limits* based on the lower levels of information.

At each level, there is incorporation of assumptions and additional information, including social, economic, and technical considerations. Each level is sequentially dependent on those preceding it. Uncertainties and assumptions present in the original data are not always explicit in the conclusions and decisions derived from them. Only the larger government agencies, companies, and unions have the resources necessary to obtain and evaluate original data directly. Others must rely on these interpretations and they may not be aware of uncertainties and assumptions. For example, much of the information in (3) and (4) consist of publications by governments, employers, and labor which propose and support particular policies or courses of action. Such information may be highly influential and yet fail to provide a balanced view or detailed argument.

There is feedback operating in which conclusions and actions at higher levels influence what data are sought, what research is done, and what scientific assumptions are used in the collection of the data. Because of the sequential effect of the uncertainties and assumptions, certain topics may be unduly magnified while others are apparently ignored.

Burger[2] and others have pointed out that much of the information used in regulation and control of chemicals does not pass through the traditional processes of scientific scrutiny and review. Few of the groups or persons affected by the information may even see it; often the data are preliminary or suggestive rather than definitive. Burger characterized these data as unmatured; detailed information which emerges later may give a different picture. Actions based on unmatured information can be justified on the basis of urgent need for health protection. However, inadequate data, especially from confidential sources, can undermine confidence if later re-evaluation finds them inaccurate.

Controversies can be the result of uncertain or incomplete information. Controversies tend to be frequent in areas like occupational cancer where workers, employers, or governments may feel that their interests could be harmed if their view did not prevail. Though controversies may surface as disputes over facts, interpretations, or predictions, they predominantly concern values. Virtually all human occupational carcinogens, as well as the policies for controlling them, have been surrounded by controversy.

Mazur[3] pointed out that technical controversies, including those surrounding carcinogens, have many elements in common. Information is usually incomplete, uncertain, and often contradictory. Incompatible assumptions by the various parties can lead to widely divergent predictions which, often, are unverifiable. A phrase commonly used in the rhetoric of controversy, "*There is no evidence* to show that ... contention or prove it wrong", may polarize and harden positions.

Decisions about occupational carcinogens must often be based on incomplete, uncertain, or suggestive data of variable quality. Yet these may be all that science can provide truthfully at a particular time. These potential problems with information can make evaluation of the impact of a decision uncertain. Uncertainty may act to magnify the perception of risk and so make a "moderate" or "feasible" course unacceptable, perhaps because the unknown is more worrisome than the known.

A report from the Science Council of Canada[4] concluded that the controversial nature of many problems encourages political rather than technical solutions. Once the scientific and technical limits of certainty have been described, the decisions are political and social. Where controversies exist, the scientific limits of uncertainty may be so broad, or the nature of the consensus may be such that purely scientific concerns can virtually be ignored.

Mazur[3] argued that technical controversy can be beneficial by bringing forward new evidence and new arguments, but it can also cause polarization, harm the reputations of participants, discredit institutions, cause further confusion and uncertainty, and delay action. Perhaps we do best not only to recognize scientific controversy as a natural part of the advance of science, but also to recognize that controversies have large, nonscientific components which should be recognized and addressed socially and politically.

Procedures have been proposed for the resolution of technical controversies, for example by separation of questions of scientific fact from questions of values. Various types of public investigations can contribute not only to public awareness, but also to assessment and determination of the issues. These forums can succeed only where the parties can find some common ground for discussion, and the scientific questions can be clearly identified. There appears to be little agreement as yet whether "science courts" or other technical forums can resolve these value-laden questions.[3-6]

CCOHS has adopted quality assurance in an attempt to counteract the problems of uncertain, controversial, or deficient information. Quality assurance is considered essential to the credibility and acceptability of its information and advice. Quality control policies of CCOHS require that all available and relevant information about carcinogens should be clearly presented in context with the strengths and limitations carefully set out, that technical language should be avoided where it obscures understanding, that the nature of any controversies and assumptions should be described well enough for them to be understood, that possible alternatives for action should be set out, and that a personal opinion may be offered where the scientist concerned has the necessary knowledge. The various points of view must be portrayed fairly. We do not know who will see the information or what purposes it may serve.

International cooperation facilitates quality assurance, enhancing the generation, accumulation, evaluation, and dissemination of trustworthy data and information. For example, the IARC monograph program, through its classification system, is influencing regulatory activity, the acceptability of certain chemicals, and the education of workers. The monographs are generally sources of reliable, evaluated information, though they are not entirely free from controversy, as the reaction to the benzene publication showed.[7] Such programs illustrate what can be accomplished when scientists work toward consensus. However, we must also recognize that *scientific consensus is not scientific proof even if it does provide a reasonable basis for action.*

II. PARTICIPATION

Participation is increasingly seen as an effective means for governments, employers, and labor to fulfill their responsibilities, individual and collective, for maintaining safe and healthy workplaces. Participation in occupational health and safety was examined by Atherley, Booth, and Kelly.[8] Their study showed that participation differed fundamentally from consultation in that it gave all the participants a voice in the decision making, which consultation did not. Consultation was the principal means of involving workers until the mid-1970s. Participation, though growing, is still uncommon.

Participation varies in level, from local to national, and in power, from advisory to regulatory. Opportunities for participation in Canada include joint health and safety committees, joint research committees, advisory councils, official inquiries, royal commissions, and government committees concerned with the preparation of acts and regulations.

Joint research committees involving government, employers, and labor, while not yet common, are becoming a necessity as research costs climb and single, unduplicated studies become increasingly important. Such studies do not fit adequately into the theoretical model of research by verification and falsification, and compensatory checks and balances should be built in. Participation can ensure the existence of such checks and balances though the value of participation does not end there.

Cooperative and coordinated research studies overseen by participation have several potential advantages. They can bring together data, facilities, experience, expertise, and resources which may not otherwise be available, and they can increase the power, quality, and acceptability (perception of balance) of research. Participation can increase the incentive to cooperate in follow-up measures; it can ensure equal access to information.

Structurally, joint research committees should have equal representation from management and workers. These committees should report regularly to management and workers, and they should approve and release all results, conclusions, and recommendations. They should appoint technical advisors for review and advice on matters such as the selection of the research team, and the design of an acceptable yet feasible research program. Staff members of CCOHS sometimes act as technical advisors for joint research projects involving the participation by workers and managers, a prerequisite for the involvement of the CCOHS staff.

Participation in research has challenges, disadvantages, and difficulties. It may require what at times seems an excessive commitment of time and resources from participants. It may require an especially strong commitment to seeing the research through to an agreed end in spite of preliminary or uncertain results which may emerge during the study. It may require a separate mechanism for the implementation of control measures based on results of the study. Participation risks the emergence of negotiated rather than scientific truths; while participation can negotiate policies and actions arising out of the scientific conclusions, it cannot reasonably negotiate the scientific conclusions, lest the scientific be subordinate to the political.

Direct participation by labor and employers in the preparation of acts and regulations is still rare in spite of some obvious potential advantages. Regulations prepared with the involvement of the representatives of those people affected can best relate to actual workplace conditions and priorities, and are most likely to gain wide acceptance.

One model for participation in regulation is the standard-setting process for setting exposure limits employed by the Swedish National Board for Occupational Safety and Health (NBOSH).[9] The first step is the preparation of documents describing the health effects of substances by a criteria group of scientific experts representing the board, employers, and labor. In the second step, the documents are passed to a regulations group, also with tripartite representation, which carries out a technical, economic, and social evaluation and recommends an exposure limit. As a final step, the recommended limit is approved or rejected by the governing board which comprises representatives of parliament, labor, employers, and the public. The process appears relatively free of conflict in that over 50 exposure limits have been set since 1978, including several for carcinogens. The permissible exposure levels generated by this process are generally lower than those prevailing in other jurisdictions, but the final determination of the effectiveness of this joint process (and all others) would be an evaluation of actual workplace conditions.

In Canada, so far, there has been some but not much joint preparation of regulations involving carcinogens, but there has been considerable decentralization because occupational carcinogens are substantially a provincial rather than a federal responsibility. Local responsibility through participation is the complement of centralized regulation. Just as centralized regulation reflects the overall social, economic, and technical priorities of society, local responsibility through participation can reflect the priorities of management and labor in individual workplaces.

Atherley[10] commented: "The question that naturally arises from the Canadian experience is whether, on a policy level, local or central decision-making is the more effective in controlling exposure to carcinogens. The question is important, and not only for Canada.

Is some sort of balance between centralized and decentralized decision-making possible? Probably not entirely, given the radically different perspectives of each bureaucracy and the people it serves. Dissonant sounds about carcinogens are harsh to the bureaucratic ear that craves administrative harmony; but to citizens in the different regions of Canada such sounds may be welcome.

No regulation, wherever the decisions are made, can be effective without a large measure of voluntary compliance at the local level; that can come about through participation. The best people to regulate a workplace are the people who work there, the management and the workers. Such people are critically dependent on a ready supply of trustworthy, complete, and intelligible information and advice on the problems of their workplace. This is where an information and advisory service can have a major impact. Employers have a responsibility to provide a safe and healthy workplace. Workers likewise have a responsibility to work in a safe and healthy manner. Regulation can reflect a failure to fulfill these responsibilities. Feelings of sovereignty on the part of management and of organized labor have in the past led to an approach in which the obligation of industry was to control hazards with little or no participation by workers, sometimes without even consultation. The development of occupational health and safety as a professional and scientific activity has in many cases also inhibited participation, because professionals may be remote from actual workplace conditions and concerns.

The advantages of local responsibility through participation include flexibility (decisions can be taken on incomplete or suggestive evidence), responsiveness to local priorities, use and development of local resources, and possibilities for experimental approaches which can be applied widely if they are successful. The disadvantages of local responsibility through participation are that it is not uniformly strong, that it depends on the financial and technical resources available to workers and management, and that it may be hindered by poor management-labor relations.

There are major requirements for effective local control of occupational (and perhaps other) cancer risks by participation. Management and workers must have (1) access to trustworthy, complete, and intelligible information about potential carcinogens (as well as other hazards), (2) education and training in the interpretation and application of the information, and (3) access to technical advice and skills in measuring and controlling exposures.

The right to refuse work involving cancer risks is available to Canadian workers in varying degrees. The strength of that right depends on the language and interpretation of the various statutes and collective agreements. At this writing, it applies only to the individual workers directly at risk and does not allow workers to act collectively on behalf of those individuals. An expansion of that right, to provide for collective action, would give workers increased power to negotiate health and safety issues in the workplace.

Local responsibility through participation is probably best carried out by joint management-labor work site health and safety committees. In most jurisdictions these committees are consultative in that they have only contributory or advisory responsibility; under these circumstances employers have a special responsibility to support the committees and to heed their advice. The effectiveness of any approach is to be judged by the acceptability of the decision process to workers and management, the stringency of the control measures, and the degree of health protection.

To judge from the Canadian experience and from that of other nations the best policy seems to be a mixture of central regulation and local responsibility through participation. Maximum permissible exposure limits specified by regulation represent the maximum levels of the carcinogen which society as a whole will tolerate; for some carcinogens, this may mean prohibition or stringent controls. All workplaces must meet these levels. Individual workplaces may better them, stimulated by participation.

Regulations, control measures, or medical procedures will not necessarily be supported or accepted simply because they are a matter of consensus among scientists and physicians. Workers, employers, and government must be confident that the actions are fair, equitable, and reflect the social priorities they place on health and safety, centrally and locally. If local responsibility by participation could be shown to be effective in reducing exposures to carcinogens and other toxic chemicals, and to be acceptable to workers and management, much expensive and frustrating regulatory effort of debatable effectiveness could be avoided.

Tripartism and other types of participation are forms of democracy and have their inherent limitations. The hard-earned experience in tripartite organizations shows that tripartism and other forms of participation can function very well as long as the limits of consensus are recognized (see Chandler[11]). Certain topics in occupational carcinogenesis may lie outside the limits of consensus. In such areas government action is necessary, but the limits may not be static; attitudes, interests and positions can change with changing circumstances and with access to new information. Participation and its exchanges of trustworthy, complete, and intelligible scientific data and information can gradually broaden the basis for consensus and create a new atmosphere of confidence. From this will grow the trust which will remain as the basis of control of occupational cancer.

REFERENCES

1. *Report of the Royal Commission on the Health and Safety of Workers in Mines*, J. M. Ham, Commissioner, Ministry of the Attorney General, Toronto, 1976.
2. **Burger, E. J.,** Unmatured science and government regulation, *J. Toxicol. Environ. Health*, 2, 389, 1976.
3. **Mazur, A.,** *The Dynamics of Technical Controversy*, Communications Press, Washington, D.C., 1981.
4. Science Council of Canada, *Regulating the Regulators: Science, Values and Decisions*, Science Council of Canada, Ottawa, 1982.
5. **Hohenemser, C. and Kasperson, J. X., Eds.,** *Risk in the Technological Society* AAAS Selected Symp. 65, Westview Press, Boulder, 1981.
6. **Raffaele, J. A.,** *The Management of Technology: Change in a Society of Organized Advocacies*, rev. ed., University Press of America, Lanham, Md., 1979.
7. Controversial cancer report: did industry exert undue pressure? *Occup. Hazards*, 43, 1982.
8. **Atherley, G. R. C., Booth, R. T., and Kelly, M. J.,** Workers' involvement in occupational health and safety in Britain, *Int. Labour Rev.*, 3, 469, 1975.
9. **Holmberg, B.,** Occupational-cancer risk management in Sweden, *Ann. N.Y. Acad. Sci.*, 363, 255, 1981.
10. **Atherley, G. R. C.,** Management of assessed risk for carcinogens in Canada: the interplay of social, political, and scientific factors, *Ann. N.Y. Acad. Sci.*, 363, 261, 1981.
11. **Chandler, G.,** NEDC at 21: the scope and limits of consensus, *R. Soc. Arts J.*, 133(5319), 134, 1983.

Chapter 8

RISK ASSESSMENT: HUMAN EXPOSURES

Emmanuel Somers

TABLE OF CONTENTS

I. INTRODUCTION

Risk assessment is all the rage: from skydiving, transport of dangerous goods, to the siting of airports and nuclear reactors the proponents of an ordered logical society have given us the hierarchical structures that can direct our decision making. The order and rationality are admirable insofar as they lead us to considered and socially acceptable solutions for, in this instance, risk in the workplace. The caveat must be, however, that societal judgments are almost invariably complex and rest on deeper emotional sources than reason. To paraphrase Nurse Edith Cavell: "rationality is not enough".

Nevertheless, the process of risk assessment through its ordered sequence of risk identification, risk estimation, risk evaluation, and finally, risk management[1] does provide a valuable technique to separate the scientific estimation from the socioeconomic evaluation that leads, in turn, to the final political judgment. The recent National Academy of Sciences report[2] on risk assessment has similarly recommended the virtue of separating the scientific bases for the decision from the regulatory mechanism. The report also suggests that as a first priority all the U.S. government regulatory agencies should adopt standard procedures for estimating the health effects of carcinogens.

Certainly risk assessment techniques aid our appraisal of the hazards in the workplace: the simplest and most direct, namely death, has been used by a number of analysts[3] to compare the risks of different occupations. As a first rough measure these data can be useful markers to areas of study and eventual regulation, although they often ignore age-adjustment and reduce complex phenomena to single-dimensional risk ranking.

It is proposed to discuss the relationship of human exposure to the risk assessment process with particular attention to the current programs in the Environmental Health Directorate and the ensuing departmental policies.

II. EXPOSURE AND SUSCEPTIBILITY

Risk is commonly defined as the product of a hazard and the probability of its occurrence. The latter, in turn, depends on both the exposure to the toxicant and the susceptibility of the human host, as shown in Figure 1. Human exposure to carcinogenic chemicals will, in turn, be determined by the sources of exposure, the possible initiating events which allow the toxicant to enter the environment, and the pathways through which individuals in the population at risk may be exposed. Knowledge of the major routes of exposure is essential, not only for an understanding of toxicological action, but also for the regulatory control of the chemical. For example, out of the occupational setting our major intake of potentially carcinogenic trihalomethanes is from drinking water,[4] as an artefact of chlorination; hence the need to establish guidelines for their presence. Once exposure has taken place, the probability of an inadventitious event occurring still depends on individual susceptibility. Most mammalian species appear to possess natural defense mechanisms that allow them to tolerate a certain level of exposure to many chemicals, and possibly to physical effects. These defense mechanisms include metabolic detoxification, immunological surveillance, tissue regeneration, and certain repair processes. Damage to a small number of liver cells, for example, may be tolerated because the normal organ functions can be maintained by the remaining tissue until regenerative cycles have replaced the damaged cells.

As examples of exposure monitoring, I should like to describe our work on a physical agent, ionizing radiation, and for chemicals outline some of our pesticide studies.

A. Ionizing Radiation

Beginning in 1951, the Department of National Health and Welfare has provided to Canadian workers the most comprehensive dosimetry service for exposure to radiation in

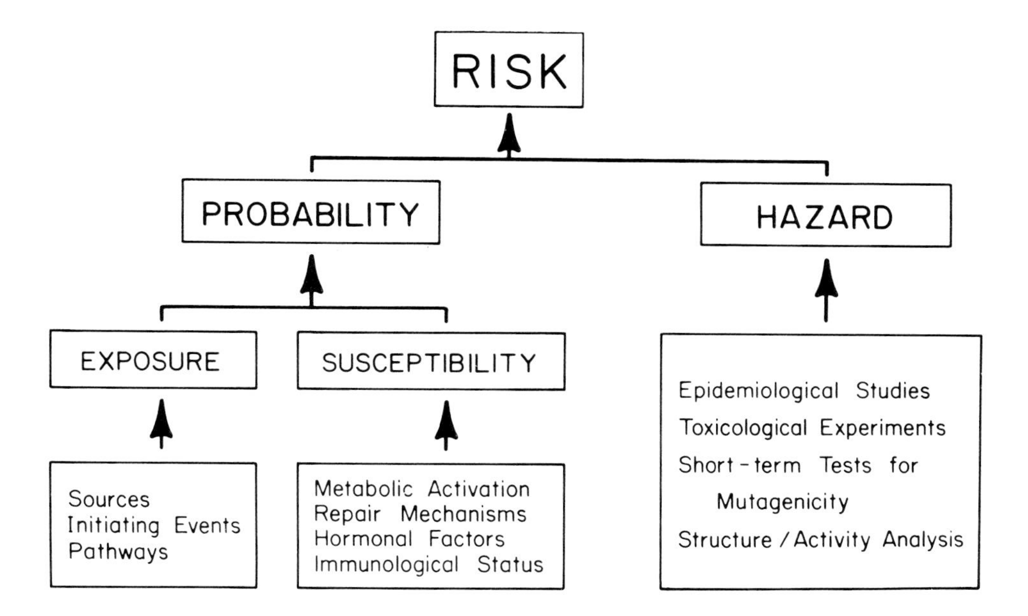

FIGURE 1. Components of risk.

the world. The initial film badge service has been replaced by thermoluminescence dosimeters and, currently, some 90,000 Canadians from 10,000 organizations are using this service for X-ray, α, γ, and neutron dosimetry. The users range from dental workers (average annual exposure 0.05 Sv), veterinarians, and industrial radiographers (average annual exposure 4.0 mSv) to university scientists. The records are computerized and form the National Dose Registry which now contains over 300,000 individual lifetime radiation exposures, some of which date back over 30 years. The dose records are unaffected by the workers changing jobs or province of employment. The Registry is primarily used to assess the long-term effects of low doses of ionizing radiation on humans and to estimate the risk of radiation exposure in the workplace: high exposures are quickly followed up and checked. It is also used for assisting the regulatory authorities in the control of occupational radiation exposures and for providing evidence of exposure to radiation for use in legal and compensation cases.

Statistical data such as average yearly doses of a variety of occupations and regions are published in an annual report on occupational radiation exposures in Canada.[5] These figures are used by various federal and provincial health and regulatory agencies to identify areas of employment where radiation hazards appear to be increasing and to show employers and the employee how their specific dose patterns compare to the provincial or national averages. The regulatory limits for atomic radiation workers are currently 30 mSv/quarter and no more than 50 mSv/year. The limit for the general public is 5 mSv/year. These limits are set by the Atomic Energy Control Board for radioisotopes and by some provinces for X-rays. The whole procedure strengthens the concept of radiation protection which requires that each employer strive to keep the doses as low as is reasonably achievable (ALARA).

Canada is unique in having this centralized radiation exposure record system as well as a National Mortality Data Base, located at Statistics Canada. In 1983, we embarked on the first phase of a large scale study, extending for 4 years, into the long-term effects of low doses of radiation on humans. It involves the computerized linkage of the National Dose Registry's radiation exposure records and the mortality records of Statistics Canada's National Mortality Data Base. Computerized record linkage has developed over the last 10 years and is proving to be an effective and economic method of linking exposure files to death files. This technique involves searching for links between pairs of records in two files by calculating, on the basis of the identification information available, the probability that a pair of records belongs to the same individual. A generalized record linkage facility now exists at Statistics Canada and is capable of handling a wide variety of applications.

Based on the current risk estimates, the size of the exposed population, and the magnitude of the exposure, the calculated number of radiation-induced cancer mortalities is very low in comparison with the number of naturally occurring cancer mortalities. A large exposed population is therefore required to confirm the current risk level of 100 radiation-induced cancer mortalities per 10^4 man-mSv. Our calculations show that (based on the growth of the nuclear industry in Canada) it will be sometime between the years 2000 and 2010 before there will be sufficient data to allow confirmation of this level of risk.

There is, however, still a great value in pursuing a study at this time. It will allow us to confirm that the current risk levels are of the right magnitude and will enable us to place an upper limit on the level of risk. It will also allow us to detect any mortalities resulting from rarer forms of cancer which may be particularly susceptible to induction by radiation.

It should be noted that for an accurate risk analysis it is necessary to convert external measures of exposure to radionuclides, but not X-rays, to internal doses at target organs and tissues. This will involve consideration of the biophysical mechanisms by which radionuclides enter the body and then migrate as for example, particles removed by the lungs ciliary clearance, or those swallowed with mucus.

B. Pesticide Exposure

The amount of pesticide to which a worker is exposed can be considered[6] in terms of three components: levels in the workplace, contact dosage, and absorbed dosage.

Levels in the workplace. Variables such as type of spray equipment, mode and rate of application, type of formulation, and duration will all affect the pesticide exposure of the worker in addition to such factors as temperature and wind velocity.

Contact dosage. The three major routes of exposure are dermal, inhalation, and to a lesser extent, oral. Although the relative importance of each route will differ the total body contact will be the sum of the amount of pesticide at each barrier (skin, lungs and gastrointestinal tract). In orchard applications it is clear that dermal exposure is the most important and absorbent patches are used to assess the concentration of pesticide on the skin although these may not always accurately reflect exposure.[7]

Absorbed dosage. The absorption of pesticide into the bloodstream will be influenced by the barrier it crosses, i.e., skin, lungs, and GI tract. Indirect indicators of absorption may be used such as the analysis of pesticide metabolites in urine; Franklin et al.[7] observed a high correlation between urinary alkyl phosphate excretion and the amount of azinphos-methyl sprayed.

Our work in monitoring air concentrations of pesticides has been aided by a new analytical technique using an atmospheric pressure chemical ionization mass spectrometer system, named Trace Atmospheric Gas Analyzer (TAGA). This instrument, designed to be mobile, can analyze the chemical directly in the air train and has been used for real-time monitoring of methomyl, at concentrations below 1 ppb, during and after spraying in a greenhouse;[8] exposure to Sulfotep® has been similarly monitored.[9] The great advantage of this system is that the pesticide can be measured under actual use conditions, rather than samples collected and measured later, so that worker exposure can be directly assessed and recommendations made for re-entry times. The TAGA can also be used for direct analysis of exhaled human breath[10] which has obvious applications for investigations in the workplace, particularly for those workers exposed to hazardous solvents.

III. HAZARD ESTIMATION

We now come to the other side of the risk equation: the estimation of hazard. The methods available to identify and estimate the risk from chemicals have been reviewed elsewhere[11] and, at the meeting out of which this chapter was promulgated, Dr. Miller has comprehensively described the use of epidemiology to uncover and measure occupational cancer. Therefore, I will concentrate on those methods that can be used to assess the carcinogenic potential of chemicals before they are introduced to the workplace, namely, toxicological experiments with animals, and short-term mutagenicity tests.

A. Toxicological Risk Estimation

The use of laboratory experiments for human risk assessment involves two critical steps: extrapolation from high to low doses and interspecies extrapolation.

The traditional approach to risk estimation is to take the highest dose inducing no effect on a sensitive animal species, the "no observed effect level" (NOEL), and use a safety factor, often 100-fold, to give the acceptable daily intake (ADI) for human populations. However, while the normal homeostatic mechanisms of the cell can provide a defense through such means as metabolic detoxification, tissue regeneration, and immunological surveillanceand threshold hypothesis has been seriously questioned for delayed, irreversible toxic effects such as carcinogenesis.[12] In addition, the use of safety factors ignores the influence of sample size and the slope of the dose-response curve.

In an attempt to overcome these difficulties, statisticians have proposed the use of mathematical modeling procedures as a means of estimating a virtually safe dose. Nevertheless, this statistical procedure is subject to certain limitations.[13] Foremost among these is the fact many different models may fit the observed data equally well in statistical terms, yet differ markedly in terms of their behavior in the low-dose region. Biologically, moreover, it is doubtful that a single mathematical model can fully describe all aspects of the mechanisms underlying a process as complex and multistage as carcinogenesis.

Arguments have been developed in favor of low-dose linearity as a conservative extrapolation,[11,14,15] but the empirical evidence is scant. Indeed, a careful re-analysis of the largest carcinogen bioassay ever conducted, the so-called mega-mouse study conducted by the U.S. National Center for Toxicological Research,[16] revealed that even data on the occurrence of liver neoplasia in over 24,000 animals are insufficient to rule out the possibility of nonlinearity in the low-dose region.[17] In a sense, linear extrapolation remains an ad hoc procedure, providing a reasonable answer to a difficult problem.

Resolution of at least part of this uncertainty may be possible by applying pharmacokinetic principles to risk extrapolation whereby the internal dose actually received by the target tissue is considered. If the pharmacokinetic processes involved in the metabolism of the carcinogen are all linear, then the administered dose or external level of exposure will be proportional to the effective dose or delivered dose at the site of action. This will not be the case, however, if any of these processes are saturable.[13] In particular, if any of the elimination or detoxification pathways are saturable, then the effective dose will increase more rapidly than the administered dose. Conversely, if activation of distributional pathways may be saturated, then the effective dose will increase at a lower rate than the administered dose.

The implications of nonlinear kinetics for low-dose risk estimations have recently been examined by Hoel and colleagues.[18] They suggest that for carcinogens that react with DNA tumor response should be related to the concentration of DNA adducts in the target organ rather than to the applied dose: many data suggest that this relationship may be linear. Thus, nonlinearity in the dose-response curve for tumor induction may be due to the kinetic processes involved in the formation of carcinogen metabolite-DNA adducts. In particular, they show in the presence of saturable elimination, detoxification, or repair processes, linear extrapolation procedures can overestimate risks in the low-dose region by several orders of magnitude.

Because of the uncertainties involved in low-dose extrapolation, several rough measures of potency have been proposed to rank chemical carcinogens according to their relative strength. Clayson[19] has proposed a potency index based on the dose that will induce an excess of 50% over spontaneous response rate. By focusing on dose corresponding to relatively high-response rates, the problem of low-dose extrapolation is circumvented. It is, however, well-established that carcinogenic potencies can differ markedly between species so that the problem of interspecies extrapolation remains.

B. Short-Term Mutagenicity Tests

These tests provide obvious advantages in economy and time and are attractive to the animal rights' groups. Chemicals that change the genetic code are often carcinogenic, although the degree of this association appears to depend greatly on the class of chemical as well as the properties of the test system used to assay mutagenic potential.[20] The tests, now up to a 100 in number, are certainly a valuable analytic tool particularly when the results can be confirmed in other short-term tests using nonrepetitive endpoints or different activation systems.[21] Directorate studies, usually with the Salmonella/mammalian-microsome assay, have found mutagenic activity in such diverse chemicals as rhodamine dyes,[22] nitroarene dyes,[23] paint removers containing dichloromethane,[24] and the herbicides diallate and triallate.[25] Certainly such results are indicative but, with the possible exception of electrophilic agents, the multistage nature of carcinogenesis is surely too complex to be reduced to elemental systems.

IV. REGULATORY ACTION ON CARCINOGENS

Regulatory strategies can be considered as ranging from an outright ban to recommendation or guidelines; some of the principles governing Canadian actions have been reviewed elsewhere.[26] Generally decisions evolve from what one trusts is an orderly evaluation of the available information. Recently, more structured decision-point approaches to carcinogenicity identification have been developed of which the most comprehensive is that of Squire[27] in which information on the nature of the neoplastic events induced in the chronic bioassay is considered along with evidence of genotoxicity. This procedure utilizes perhaps the most relevant toxicological data to construct five different classes of animal carcinogens, thereby providing a possible basis for prioritizing regulatory concerns. There is no doubt but that these structured approaches to hazard identification demand a systematic analysis of the toxicological data in a rational fashion. However, checklists do not make decisions but can best serve as a guide to the decision maker.

In Canada, the majority of the workforce, at least 90% and hence the majority of the occupational health legislation lies in the provincial, not federal, domain. Nevertheless, there are a number of regulatory actions taken at the federal level that illustrate the use of risk assessment procedures for industrial carcinogens.

A. Nitrosamines

Synthetic cutting fluids used to reduce friction during metal grinding can contain ethanolamines as emulsifiers and nitrite as a corrosion inhibitor. These components can react to form the carcinogenic nitrosamines. Our analyses[28] showed that 8 out of 24 samples of cutting oils contained *N*-nitrosodiethanolamine at concentrations up to 0.5%. Although the extent of worker exposure was not known the level of the presumed risk was considered serious enough to ban, under the Hazardous Products Act, the sale of cutting oils containing both nitrites and ethanolamines.

B. Formaldehyde

The level of formaldehyde is, of course, controlled in the workplace but the action of the federal government in banning the sale of urea-formaldehyde foam insulation (UFFI) under the Hazardous Products Act in 1980 was largely based on the toxic, particularly carcinogenic, properties of formaldehyde. The acute, irritating properties of formaldehyde gas are well established. Rodent studies have shown that formaldehyde can induce nasal carcinomas at relatively low concentrations. When subjected to statistical analysis using a linearized multistage model, these studies suggested that exposure to 0.04 ppm formaldehyde over and above ambient levels might produce an additional 32 cancers in a population of size 1 million over a 16-hr/day exposure for 6 years.[29] In fact, we used this figure to estimate that the recommended indoor air guideline of 0.1 ppm formaldehyde would give a cancer risk possibly less than that of smoking one third of a cigarette each day. These are the health risks associated with formaldehyde, but one must also recognize that knowledge of its insulation inadequacies and adverse effects on building structure as well as the availability of safer, better insulation materials contributed to the ultimate political decision to ban.

C. Asbestos

The most notorious physical carcinogenic agent is asbestos and Canada is the second major producer in the world with some 35,000 workers employed directly, or indirectly, in the industry. In 1975, The Minister of National Health and Welfare recommended to his federal and provincial colleagues that the workplace standard should be no more than 2 fibers per cubic centimeter, time-weighted average, length greater than 5 μm. This value was adopted by the provinces and, in some cases, reduced and strengthened. The use of asbestos in children's toys, modeling clays, drywall cements, or simulated ashes has been banned, through the sale of these products, under the Hazardous Products Act.

Although reduction of the exposure of workers almost certainly lessens the risk of all asbestos-related disease, the 2 fibers per cubic centimeter value was based on an acceptable level of risk for the development of asbestosis. At the present time, available epidemiological data are insufficient to permit a precise determination of the cancer risks associated with exposure to asbestos concentrations in the workplace at or below these levels, and there is considerable variation in estimates of this risk which have been derived through linear extrapolation of dose-response data from historical prospective mortality studies of occupationally exposed populations. It has been estimated that a 1% excess in lung cancer mortality is associated with lifetime occupational exposure to asbestos concentrations of 0.06 to 1.25 fibers per cubic centimeter;[30] the range of this estimate is partially due to problems inherent in using retrospective epidemiological studies for risk extrapolation and partially to the complex nature of asbestos. The 2 fiber per cubic centimeter level is currently being reviewed by the department.

D. Polychlorinated Biphenyls

The polychlorinated biphenyls (PCBs) are regarded as persistent and ubiquitous environmental contaminants with carcinogenic potential. Although the restrictive legislation was directed to environmental concerns, it also clearly affected workplace safety as well. Under the Environmental Contaminants Act the uses of PCBs were limited in 1977 to closed electrical and heat-transfer systems. Subsequently, in 1980 all new uses, including use in makeup or filling fluid, were banned. In view of recent concerns of the presence of low concentrations of PCBs in the office environment we have supported the proposed NIOSH limit of 1 $\mu g/m^3$. Our estimates, using linear extrapolation from the rat liver tumors study with Aroclor 1260,[31] is that this level would present an additional lifetime risk of cancer of less than 1 in 10^5.

V. CONCLUSION

It should be apparent that the formal procedures of risk estimation are only as powerful as their scientific base. We use them to aid the final societal judgment in what we trust is a measured and responsible fashion. Nevertheless, it is only through greater scientific knowledge, particularly of the biological mechanisms of toxicity and assessment of target exposure, that these judgments will be improved and strengthened.

REFERENCES

1. **Whyte, A. V. and Burton, I., Eds.,** *Environmental Risk Assessment,* Scope 15, Wiley, New York, 1980, chap. 1.
2. Committee on the Institutional Means for Assessment of Risks to Public Health, *Risk Assessment in the Federal Government: Managing the Process,* National Academy Press, Washington, D.C., 1983.
3. **Pochin, E. E.,** Occupational and other fatality rates, *Comm. Health,* 6(2), 2, 1974.
4. **Hickman, J. R., McBain, D. C., and Armstrong, V. C.,** The contribution of drinking water to exposure to toxic substances in Canada, *Environ. Monitoring Assessment,* 2, 71, 1982.
5. Environmental Health Directorate, Health Protection Branch, *Occupational Radiation Exposures in Canada — 1980,* 82-EHD-79, Department of National Health and Welfare, Ottawa, 1981.
6. **Franklin, C. A., Muir, N. I., and Greenhalgh, R.,** The assessment of potential health hazards to orchardists spraying pesticides, in *Pesticide Residues and Exposure,* ACS Symp. Ser. 182, Plummer, J. R., Ed., American Chemical Society, Washington, D.C., 1982, 157.
7. **Franklin, C. A., Fenske, R. A., Greenhalgh, R., Mathieu, L., Denley, H. V., Leffingwell, J. T., and Spear, R. C.,** Correlation of urinary pesticide metabolite excretion with estimated dermal contact in the course of occupational exposure to Guthion, *Toxicol. Environ. Health,* 7, 715, 1981.
8. **Williams, D. T., Denley, H. V., Lane, D. A., and Quan, E. S. K.,** Real time monitoring of methonyl air levels during and after spraying in a greenhouse, *Am. Ind. Hyg. Assoc. J.,* 43, 190, 1982.
9. **Williams, D. T., Denley, H. V., and Lane, D. A.,** On-site determination of Sulfotep air levels in a fumigating greenhouse, *Am. Ind. Hyg. Assoc. J.,* 41, 647, 1980.
10. **Benoit, F., Davidson, W. R., Lovett, A. M., Nacson, S., and Ngo, A.,** Breath analysis by atmospheric pressure ionization mass spectrometry, *Anal. Chem.,* 55, 805, 1983.
11. **Somers, E. and Krewski, D.,** Risks from environmental chemicals, in *Proc. Symp. Assessment and Perception of Risk to Human Health in Canada,* Rogers, J. T. and Bates, D. V., Eds., Royal Society of Canada, Ottawa, 1983, 43.
12. **Schneiderman, M. A., Decouflé, P., and Brown, C. C.,** Thresholds for environmental cancer: biologic and statistical considerations, *Ann. N.Y. Acad. Sci.,* 329, 92, 1979.
13. **Krewski, D. R., Clayson, D., Collins, B., and Munro, I. C.,** Toxicological procedures for assessing the carcinogenic potential of agricultural chemicals, in *Genetic Toxicology: An Agricultural Perspective,* Fleck, R. A., Ed., Plenum Press, New York, 461, 1982.
14. **Crump, K. S.,** Statistical aspects of linear extrapolation, in *Health Risk Analysis,* Richmond, C. R., Walsh, P. J., and Copenhaver, E. D., Eds., Franklin Institute Press, Philadelphia, 1981, 381.
15. Environmental Protection Agency, Water quality criteria: availability, *Fed. Regist.,* 44, 56627, 1979.
16. **Stafford, J. A. and Mehlman, M. M.,** *Toxicants and Drugs: Kinetics and Dynamics,* Wiley, New York, 1979.
17. **Brown, K. and Hoel, D.,** Modelling time-to-tumor data: analysis of the ED_{01} study, *Fundam. Appl. Toxicol.,* 3, 458, 1983.
18. **Hoel, D. G., Kaplan, N. L., and Anderson, M. W.,** Implication of nonlinear kinetics on risk estimation in carcinogenesis, *Science,* 219, 1032, 1983.
19. **Clayson, D.,** Problems in interspecies extrapolation, in *Toxicological Risk Assessment,* Clayson, D., Krewski, D., and Munro, I., Eds., CRC Press, Boca Raton, Fla., in press.

20. **Purchase, I. F. H.,** Quantitative estimation of the risks of human health: a comparison of the potency of mutagenic effects of chemicals in short-term tests with their effect in rodent carcinogenicitis experiments, *Proc. Workshop on Quantitative Estimation of Risk to Human Health from Chemicals,* in press.
21. **Bartsch, H., Tomatis, L., and Malaveille, C.,** in *Regulatory Toxicol. Pharmacol.,* 2, 94, 1982.
22. **Nestmann, E. R., Douglas, G. R., Matula, T. I., Grant, C. E., and Kowbel, D. J.,** Mutagenic activity of Rhodamine dyes and their impurities as detected by mutation induction in *Salmonella* and DNA damage in Chinese hamster ovary cells, *Cancer Res.,* 39, 4412, 1979.
23. **Nestmann, E. R., Kowbel, D. J., and Wheat, J. A.,** Mutagenicity in *Salmonella* of dyes used by defence personnel for the detection of liquid chemical warfare agents, *Carcinogenesis,* 2, 879, 1981.
24. **Nestmann, E. R., Otson, R., Williams, D. T., and Kowbel, D. J.,** Mutagenicity of paint removers containing dichloromethane, *Cancer Lett.,* 11, 295, 1981.
25. **Douglas, G. R., Nestmann, E. R., Grant, C. E., Bell, R. D. L., Wytsma, J. M., and Kowbel, D. J.,** Mutagenic activity of diallate and triallate determined by a battery of *in vitro* mammalian and microbial tests, *Mutat. Res.,* 85, 45, 1981.
26. **Somers, E.,** Environmental risk management in Canada, *Regulatory Toxicol. Pharmacol.,* 3, 75, 1983.
27. **Squire, R. A.,** Ranking animal carcinogens: a proposed regulatory approach, *Science,* 214, 877, 1981.
28. **Williams, D. T., Benoit, F., and Muzika, K.,** The determination of N-nitrosodiethanolamine in cutting fluids, *Bull. Environ. Contam. Toxicol.,* 20, 206, 1978.
29. Consumer Product Safety Commission, *Fed. Regist.,* 46, 11187, 1981.
30. **Meek, M. E., Shannon, H., and Toft, P.,** Asbestos case study, in *Toxicological Risk Assessment,* Clayson, D., Krewski, D., and Munro, I., Eds., CRC Press, Boca Raton, Fla., in press.
31. **Kimbrough, R. D., Squire, R. A., Linder, R. E., Strandberg, J. D., Montali, R. J., and Burse, V. W.,** Induction of liver tumors in Sherman strain female rats by polychlorinated biphenyl Aroclor 1260, *J. Natl. Cancer Inst.,* 55, 1453, 1975.

Chapter 9

VINYL CHLORIDE: A CASE HISTORY OF REGULATORY ACTION IN RELATION TO SCIENTIFIC KNOWLEDGE OF CANCER-CAUSING EFFECTS*

Peter F. Infante

TABLE OF CONTENTS

* The views expressed do not necessarily represent those of the Occupational Safety and Health Administration.

I. EVENTS LEADING TO AN EMERGENCY TEMPORARY STANDARD

In 1926, vinyl chloride (VC) was introduced into American commerce. VC was shown to cause cardiac arrhythmia in experimental animals by 1947, and thus was discontinued for use as a general anesthetic agent. VC was first reported to be associated with hepatic abnormalities in workers by 1949, was associated with toxic angioneuropathy among workers exposed below the Russian maximum allowable concentration of 400 ppm by 1957, was reported to cause centrilobular granular degeneration of the liver at 200 ppm exposure by 1961, and was shown to induce acro-osteolysis in workers cleaning reactor vessels by 1966.

In 1971, Viola and others[1] reported the induction of tumors of the skin, lung, and bones in rats exposed by inhalation to 30,000 ppm of VC. These oncogenic findings were reported to have elicited little response in 1971 because the testing results were available from studies conducted only at unrealistically high doses bordering on the lower explosive limit of VC. In reality, however, this was not the case. As early as May of 1971, Dr. Viola had presented several U.S. companies with unpublished findings interpreted by industry scientists as demonstration of an increased incidence of tumors in rats exposed to VC concentrations down to, and including, 5000 ppm.[2] Additional ongoing studies involving VC exposures of 20,000 ppm to less than 500 ppm were also described to those same companies. In like manner, Maltoni in August 1972 reported to the Manufacturing Chemists Association and to several chemical companies in Europe findings of liver angiosarcoma and cancers of other sites in rats exposed by inhalation to VC at lower concentrations.[14] As late as July 1973 when representatives of industry met with NIOSH "the presentation was to be oral with only copies of the study protocols and generally-known data provided for distribution. The objectives were to apprise NIOSH of vinyl chloride studies in progress and to avoid over-reaction by the Department of Labor and NIOSH should data become available from this or other sources...Needless to say, any breach of the presentation's informal confidentiality could be extremely costly to industry."[3] Government, labor, and the independent research community were first informed of the expanding knowledge of the carcinogenicity of VC in January 1974, when representatives of industry announced that they had found liver angiosarcoma in two workers who cleaned reactor vessels as part of their employment at the B.F. Goodrich plant in Louisville, Ky. Almost simultaneously, the findings of the induction of liver angiosarcoma in rats exposed to VC were made public.

The polymerization workers who died from angiosarcoma entered reactor vessels as part of their job. The workers would steam out one vessel and then go inside and chip off the remaining partially polymerized polyvinyl chloride (PVC). They would then move on to the next vessel and eventually clean out an entire series.

Under Section 6(c) of the Occupational Safety and Health Act, the Secretary of Labor may issue an emergency temporary standard to take immediate effect upon publication in the *Federal Register* if he determines (a) that employees are exposed to grave danger from exposure to substances or agents determined to be toxic or physically harmful or from new hazards, and (b) that such an emergency standard is necessary to protect employees from such danger. Thus, in April 1974, OSHA promulgated an emergency temporary standard (ETS) of 50 ppm VC in the air as a ceiling concentration (the standard up to that time had been a 500 ppm ceiling level). The issuing of an ETS must also be followed by the issuing of a permanent standard within 6 months of publication of the ETS. Thus, the permissible exposure limit was reduced from a ceiling concentration of 500 ppm to 50 ppm on an emergency basis; 1 month later, in May 1974, OSHA proposed a 1 ppm, 8-hr time-weighted average (TWA) permissible exposure level. This standard requires atmospheric monitoring for VC. The standard also affords employees the opportunity to observe the monitoring and the resulting measurements. It includes engineering controls and specific work practice measures and furthermore, it requires that engineering controls be the sole means by which this reduction in exposure is achieved. Respirators are made available and are to be used only for compliance in an emergency situation. As part of the standard, worker education about the hazard, protective clothing, industrial hygiene facilities and practices, and medical surveillance, including tests to detect liver dysfunction are provided. A final standard was issued in October 1974.

A. Experimental Evidence Available Prior to Promulgation of 1974 OSHA Standard

Information presented by Maltoni at a February 1974 OSHA hearing[4] demonstrated that exposure to VC by inhalation at and below the then permissible exposure limit of 500 ppm induced tumors, including angiosarcoma of the liver in three species of animals. Tumors had been observed in animals exposed to VC at concentrations as low as 250 ppm. No tumors had yet been observed by Maltoni in animals exposed to 50 ppm VC. Furthermore, data reported by Torkelson and others,[4] indicated that exposure to VC at concentrations of 50 ppm failed to induce tumors in rats, hamsters, rabbits, and dogs. Although only preliminary in nature, a study sponsored by the Manufacturing Chemists Association (MCA) revealed that 2 out of 200 mice exposed to VC concentrations of 50 ppm developed angiosarcoma of the liver. On the basis of this information, OSHA concluded that the 50 ppm level established in the ETS issued in April was inadequate and proposed in May that exposures should be lowered to the lowest detectable level, which at the time was 1 ppm. Prior to issuing a final standard in October 1974, additional experimental evidence by Maltoni[5] indicated the induction of cancers of the liver (angiosarcoma), kidneys, lungs, and skin as a result of 50 ppm VC exposure. The MCA study also demonstrated cancers of multiple sites associated with 50 ppm VC exposure.

B. Epidemiological Evidence Available Prior to Promulgation of 1974 OSHA Standard

Employees of the B.F. Goodrich plant mentioned earlier, who died from angiosarcoma of the liver, had an average exposure of 19 years to unknown concentrations of VC and variable exposures to other volatile chemicals.[5] Employees of three other companies were also reported to have had exposure to VC and to have died from angiosarcoma of the liver.[5] By August 1974, 21 cases of liver angiosarcoma (13 from the U.S., plus 7 from other countries) had been identified among VC polymerization workers.[6] Early epidemiologic observations by NIOSH,[7] Mt. Sinai,[8] and Tabershaw and Gaffey[9] indicated an increase in cancers of the liver, lung, brain, and lymphohematopoietic system among workers exposed to high levels of VC. (The reports of the final results of these studies are listed among the references.)

C. The Unreported Canadian Experience

The U.S. government was not aware of the Canadian experience at the time of the 1974 OSHA hearings. It was not until much later that we were aware of it.

To my knowledge, the earliest death from angiosarcoma of the liver that the U.S. government eventually became aware of among PVC workers, was that of a Canadian worker on September 2, 1955. A second death from angiosarcoma of the liver occurred among workers at the same Canadian plant on December 21, 1957. Before the health community became aware of the possibility of a serious health hazard related to VC exposure on January 22, 1974, eight cases of angiosarcoma of the liver had already been seen at this plant.[10] As the late Dr. J. William Lloyd, one of the major NIOSH scientists involved in the VC episode stated:[11] "I would suggest to you that the frequency of such a rare disease in a plant employing at most a few thousand workers over the years and perhaps only several hundreds, should have become apparent to even an unsophisticated management, and most certainly would have been obvious to company medical personnel or medical consultants. In addition to the high liver cancer mortality of the Canadian workers, the unusual age distribution should have alerted management and medical personnel to a very unusual problem. The first case died at age 41. The second at age 43. The third at age 42. Of the 10 deaths from that plant to date, 8 died at age 53 or younger. The oldest occurred at age 61.''

D. Interpretation of Evidence Available to OSHA In 1974

The U.S. Surgeon General's Ad Hoc Committee on the Evaluation of Low Levels of Environmental Chemical Carcinogens in 1970 concluded that safe exposure levels for carcinogenic substances cannot be scientifically determined. This position was supported by testimony at the 1974 OSHA hearings by expert witnesses from NIOSH and from the U.S. National Cancer Institute. They concluded there was no way that safe levels for carcinogens could be determined. The Surgeon General's committee also concluded that the finding of cancer in two or more animal species may be extrapolated to indicate a carcinogenic hazard to humans. In the VC situation, a finding was made in three species exposed by inhalation — a route comparable to human exposure.[5] There were, however, some witnesses who suggested that humans are less sensitive to the carcinogenic effects of VC than experimental animals. Proponents of this position argued that if humans were as sensitive as rodents, an epidemic of cancer resulting from VC should have already been discovered among employees.[5] However, not a single epidemiological study had been completed by the time of the 1974 hearings and as preliminary results began to unfold, an epidemic was apparent. Reports of more than 20 cases of liver angiosarcoma, an extreme rarity in the U.S., added to this concern.

It was known that the polymerization workers were exposed to high levels of VC. When workers climbed into the reactor vessels to clean them out, it was considered in some plants a good industrial hygiene practice to tie a rope around their belt so they would not fall to the bottom of the reactor vessel and have a serious injury if they passed out from the anesthetic effects of VC. Since employees in whom tumors had been observed were those who had had considerable employment experience in polymerization operations where exposure to VC had been high, it was argued that the lower levels found in the workplace in 1974, roughly between 50 and 500 ppm, could not induce cancer and therefore were safe. This position was purported to be supported by a study of 335 DOW workers, exposed to VC levels of less than 200 ppm, who did not exhibit an excess of cancer over a 7-year monitoring period. Because of methodological deficiencies in this study and experimental bioassay results demonstrating cancer at 50 ppm VC exposure, this argument was rejected. OSHA concluded that demonstration of cancer induction in humans at a particular level is not a prerequisite to a determination that a substance represents a cancer hazard for humans at that level. Furthermore, animal tests of VC exposures below 50 ppm had not yet been completed and OSHA concluded it would have been unfounded to assume that animals exposed to VC concentrations below 50 ppm would not develop tumors.[5]

II. CANCER INFORMATION DEVELOPED SUBSEQUENT TO THE 1974 OSHA STANDARD

A. Experimental Evidence for Carcinogenicity of Vinyl Chloride

Since 1974, there have been numerous studies of the experimental carcinogenicity of VC. (For a review of this information through 1978, see IARC[12] Vol. 19, 1979; and more recently *Environmental Health Perspectives,*[13] Vol. 41, 1981.) The final results of the series of studies in the rat, mouse, and hamster by Maltoni and co-workers[14] demonstrate the induction of liver angiosarcoma, hepatoma, tumors of the brain and lung, lymphomas and leukemias, angiosarcoma and angiomas of ''other sites,'' nephroblastomas, sebaceous cutaneous carcinomas, other cutaneous epithelial tumors, forestomach tumors, melanomas, and mammary carcinomas. With regard to exposure level, liver angiosarcoma was observed down to the level of 25 ppm, Zymbal gland carcinomas down to a level of 10 ppm, and mammary carcinomas down to the level of 5 ppm.[14] A study on the effect of ethyl alcohol administered in the drinking water of rats exposed by inhalation to VC has shown more than a doubling of the incidence of liver angiosarcoma in these animals as compared to those receiving VC alone.[15] Thus, alcohol enhances the carcinogenic effect of VC.

B. Epidemiologic Evidence for the Carcinogenicity of Vinyl Chloride

1. Occupational Exposure to Vinyl Chloride and Cancer

The final results of the epidemiological studies of workers exposed to VC were reviewed in 1979 by IARC[12] and more recently by Infante.[16] The studies demonstrate a significant excess of cancer of the liver, brain, and lung and suggest an excess of lymphatic and hematopoietic system cancer deaths. Exposure levels associated with the excessive cancer risk to the VC polymerization workers is unknown.

What is known about low level occupational exposure to VC and cancer? To answer this question it is necessary to review the results of studies of individuals employed in the PVC fabricating industry. Exposure to VC among PVC fabricators ranged between 0 and 15 ppm. Typical fabrication products include coated wire, upholstery fabrics, floor and wall coverings, pipe and other construction materials, toys, recreational equipment, phonograph records, containers and container lining, and a myriad of novelty items. Because they were fabricating PVC products, their only exposure to VC would be through the residual VC monomer that was trapped in the PVC they happened to be molding. A study[17] of 3847 deaths has demonstrated significantly elevated proportional mortality for "total cancer" and for cancers of the digestive system among both male and female employees, and for cancer of the breast among female employees in the PVC fabricating industry. A subsequent case-control analysis for the breast cancer deaths did not reveal any exposure significantly associated with breast cancer among these female employees. However, the authors noted that the smallest relative risk that could have been detected was nearly 3:1. Therefore, in relation to the excessive risk of breast cancer observed in the PMR study (37% excess, $p < 0.05$), the case-control study lacked the sensitivity to evaluate whether this elevated risk was related to low-level VC exposure. These findings are of interest in light of results demonstrating the induction of mammary carcinoma in rats exposed to 5 ppm VC.[14]

With further regard to low-level exposure to VC and cancer, Christine et al.[18] reported two histopathologically confirmed angiosarcoma cases identified among employees in facilities located in Connecticut where employees fabricated vinyl sheets and also coated wire with PVC. One of these individuals, a 47-year-old man, had worked for the previous 10 years as an accountant in a factory producing vinyl sheets and processing PVC resins. This individual had frequently visited the plant's production area. The second individual, a 61-year-old man, had spent 25 years in an electrical plant operating a machine that applied PVC-containing plastic to wires. Baxter et al.[19] in a review of 14 cases of hepatic angiosarcoma diagnosed in Great Britain during 1963 to 1973, noted one case who had worked with a process which used PVC as a raw material.

2. Community Exposure to Vinyl Chloride

To further address the carcinogenic effects from low level VC exposure, let us look at what we know about community exposure to VC. In order to evaluate any effect from such exposures, it would seem to be most reasonable to study the occurrence of angiosarcoma of the liver because of its rarity and its known association with VC.

Christine et al.[18] reported two cases of hepatic angiosarcoma in Connecticut having a probable residential exposure to VC. One individual with hepatic angiosarcoma lived within 2 mi of the plant producing PVC-coated wire, while the second individual lived within 0.5 mi of a plant producing vinyl sheets. Each of these two plants also had a case of hepatic angiosarcoma among its labor force. Neither of these residential cases was known to have had occupational exposure to VC or arsenic or diagnostic exposure to thorium dioxide, the only three agents known to cause hepatic angiosarcoma in humans. Baxter et al.[19] reported the diagnosis of another case of liver angiosarcoma in Great Britain in an individual who lived for 6 years within 0.5 mi of a plant manufacturing PVC. On the basis of these observations, Brady et al.[20] in 1977 undertook a study in New York State of 26 confirmed cases of hepatic angiosarcoma. Controls, comprised of individuals who had an internal malignant tumor other than primary liver cancer, were matched with index cases on the basis of age at diagnosis, race, sex, place of residence, and vital status. This study showed a statistically significant association between angiosarcoma of the liver and direct occupational or therapeutic exposure to arsenic (2 cases), VC (3 cases), and thorium dioxide (2 cases). In addition, this study demonstrated that of 10 female cases of liver angiosarcoma (no direct occupational exposure to VC or arsenic, or therapeutic exposure to thorium dioxide) one lived within 1700 feet of a VC polymerization plant while four additional women lived from 500 to 4500 feet of a PVC fabrication plant. In contrast, none of their matched controls lived within 1 mi of any facility polymerizing VC or fabricating PVC. These study findings are supportive of the role of indirect and low level VC exposure in the etiology of cancer in humans.

III. CONCLUSIONS

Data presented in a scientific forum in 1970 demonstrated that exposure of rats to 30,000 ppm vinyl chloride induced cancer. Because of the high dose administered these findings unfortunately were ignored by the government. Study results of the carcinogenic effects from exposure to 5000 ppm vinyl chloride were presented to American manufacturers in early 1971. These and additional experimental study results were not transmitted to the government until early 1974. In January 1974, the government was first notified of two deaths from angiosarcoma of the liver that occurred between September 1971 and December 1973 among employees of a single U.S. facility. On the basis of this information in April 1974, OSHA promulgated an emergency temporary standard (ETS) allowing a 50 ppm average exposure level. By October 1974, OSHA issued its permanent standard allowing an 8-hr TWA exposure of 1 ppm.

Contrary to industry opinion prior to the release of scientific data demonstrating the carcinogenicity of VC in experimental animals, the U.S. Department of Labor and NIOSH did not overreact to the new information on VC. The new standard was achieved easily and was beneficial to both management and workers. The VC standard did not result in plant closings, it did not result in thousands of workers becoming unemployed, and it did not result in the price of plastics skyrocketing as contended at the OSHA hearing. The new technology was cost effective in that it resulted in the companies retaining vast amounts of VC that had previously been lost through fugitive emissions.

Had it not been for the identification of VC-induced cancers in workers, a standard probably would not have been promulgated in such an expeditious manner. Furthermore, had the initial cancers observed in workers not been of a rare type, the carcinogenicity of VC in humans may have gone undetected. The occurrence of eight deaths from angiosarcoma of the liver by 1973 among individuals employed at a single Canadian facility does not leave one with much confidence in the current practice of epidemiological cancer surveillance. Furthermore, it demonstrates that one can rarely rely upon medical reports of cancer to detect workplace epidemics. VC serves as an example of good public health practice with the data available at the time. It does not serve as the usual case of "paralysis-by-analysis." VC also serves as an example of the extension of the hazard from the workplace into the community environment. Finally, VC serves as an example of the utility of laboratory bioassay in the regulation of carcinogens. For the future, it also serves as a lesson in the peril of ignoring experimental evidence of carcinogenicity.

REFERENCES

1. **Viola, P. L., Bigotti, A., and Caputo, A.,** Oncogenic response of rat skin, lungs, and bones to vinyl chloride, *Cancer Res.*, 31, 516, 1971.
2. **Wheeler, R. N.,** Union Carbide Corporation, Manufacturing Chemists Association Occupational Health Committee-Vinyl Chloride Conference: memo to several company representatives from chairman of the Occupational Health Committee of MCA, November 23, 1971.
3. **Wheeler, R. N.,** Union Carbide Corporation internal correspondence, Vinyl chloride research: MCA report to NIOSH, July 19, 1973.
4. Occupational Safety and Health Administration, Department of Labor (29 CFR 1910), Vinyl Chloride: proposed standard, May 10, 1974.
5. Occupational Safety and Health Administration, Department of Labor (29 CFR 1910) Standard for Exposure to Vinyl Chloride, October 4, 1974.
6. **Wagoner, J. K.,** Statement before the Subcommittee on the Environment of the U.S. Senate Commerce Committee, August 1974.
7. **Waxweiler, R. J., Stringer, W., Wagoner, J. K., Jones, J., Falk, H., and Carter, C.,** Neoplastic risk among workers exposed to vinyl chloride, *Ann. N.Y. Acad. Sci.*, 271, 40, 1976.
8. **Nicholson, W. J., Hammond, E. D., Seidman, H., and Selikoff, I. J.,** Mortality experience of a cohort of vinyl chloride-polyvinyl chloride workers, *Ann. N.Y. Acad. Sci.*, 246, 225, 1975.
9. **Tabershaw, I. R. and Gaffey, W. R.,** Mortality study of workers in the manufacture of vinyl chloride and its polymers, *J. Occup. Med.*, 16, 509, 1974.
10. **Delorme, F. and Theriault, G.,** Ten cases of angiosarcoma of the liver in Shawinigan, Quebec, *J. Occup. Med.*, 20, 338, 1978.
11. **Lloyd, J. W.,** The fallacies of acceptable risk, in *Risk/Benefit Decisions and the Public Health: Proc. 3rd FDA Sci. Symp.*, Staffa, J. A., Ed., Office of Health Affairs, Food and Drug Administration, Washington, D.C., 1978.
12. International Agency for Research on Cancer, IARC Monographs on the Evaluation of the Carcinogenic Risk of Chemicals to Humans, Vol. 19, IARC, Lyon, 1979.
13. National Institute of Environmental Health Sciences, National Institute for Occupational Safety and Health, and Occupational Safety and Health Administration, Conference to reevaluate the toxicity of vinyl chloride monomer, poly(vinyl chloride) and structural analogs, *Environ. Health Perpect.*, 41, 1, 1981.
14. **Maltoni, C., Lefemine, G., Ciliberti, A., Cotti, G., and Carretti, D.,** Carcinogenicity bioassays of vinyl chloride monomer: a model of risk assessment on an experimental basis, *Environ. Health Perspect.*, 41, 3, 1981.
15. **Radike, M. J., Stemmer, K. L., and Bingham, E.,** Effect of ethanol on vinyl chloride carcinogenesis, *Environ. Health Perspect.*, 41, 59, 1981.
16. **Infante, P. F.,** Observations of the site-specific carcinogenicity of vinyl chloride to humans, *Environ. Health Perspect.*, 41, 89, 1981.
17. **Chiazze, L. and Ference, L. D.,** Mortality among PVC-fabricating employees, *Environ. Health Perspect.*, 41, 137, 1981.
18. **Christine, B. W., Barrett, H. S., and Lloyd, D. S.,** Angiosarcoma of the liver-Connecticut, *Morb. Mort. Wkly. Rep.*, 23, 210, 1974.
19. **Baxter, P. J., Anthony, P. P., MacSween, R. N. M., and Scheuer, P. J.,** Angiosarcoma of the liver in Great Britain, 1963—73, *Br. Med. J.*, 2, 919, 1977.
20. **Brady, J., Liberatore, F., Harper, P., Greenwald, P., Burnett, W., Davies, J. N. P., Bishop, M., Polan, A., and Vianna, N.,** Angiosarcoma of the liver: an epidemiologic survey, *J. Natl. Cancer Inst.*, 59, 1383, 1977.

Part C
Exposed Populations

Chapter 10

ISSUES IN MONITORING POPULATION EXPOSURES

Jonathan B. Ward, Jr.

TABLE OF CONTENTS

I. INTRODUCTION

The risks associated with exposure to carcinogenic chemicals have become a growing source of public concern. Steadily increasing numbers of agents are being identified as mutagens and carcinogens. Furthermore, we are increasingly aware of sources of exposure to such agents both in the occupational setting and general environment. There is a growing need for reliable and sensitive methods for detecting human exposure to carcinogenic hazards at the time of their occurrence. By detecting hazards early, significant reduction in the risk of adverse health outcome can be achieved by reduction or termination of exposure.

The identification of carcinogenic agents and assessment of risk related to exposure to them is based on the use of two methodologies: (1) in vitro and animal studies in the laboratory and (2) epidemiological studies in human populations. The detection of human exposures to biologically significant levels of carcinogenic agents is a complex matter which neither of these approaches really addresses.

Although laboratory studies can identify hazards, quantitate dose response effects, and explore mechanisms of action, extrapolation of results to human risk is often questioned. In addition to uncertainties about the similarities of human and animal response, the complexities of human exposure and variability of the human genetic background cannot be duplicated in the laboratory.

Epidemiological studies do reflect human experience, but for use in measuring carcinogenic risk, studies require large populations with well-documented histories of exposure and health outcome which are rarely available. Even in those circumstances where good studies can be performed they only identify the hazard after the health of the studied population has been adversely affected. The opportunity for prevention has been missed. Populations experiencing current exposures may face conditions very different from those of earlier studied populations in terms of level of exposure and concomitant exposures so that risk assessment again requires extrapolation.

Both laboratory and epidemiological studies are vital to the detection of carcinogenic chemicals and the evaluation of their risk to man. However, neither approach can identify human exposures to carcinogens as they occur. Reliance on measurement of environmental levels of carcinogens is not a satisfactory means to detect exposure for several reasons. A complete ascertainment of all the hazardous, or potentially hazardous, chemicals in an environment is usually not available. Even if the levels of agents of concern are known, workers may move about the environment in a complex manner which is difficult to document. Integrating the total exposure then becomes difficult. Finally additional exposures due to diet, medications, lifestyle factors, or home environment may interact with occupational exposures modifying responses to agents in ways not predicted by animal tests.

For these reasons methodologies for detecting exposures to genetically hazardous chemicals should include tests which measure biological responses to exposure. In this way all factors contributing to exposure can be integrated. This paper will discuss the use of biological monitoring tests for the detection of human population exposures to genetically toxic agents (including carcinogens), the rationale on which such tests are based, and the characteristics of some of those tests. Issues in the design and interpretation of studies using batteries of tests will be considered and, as an example, a recently conducted study of autopsy service workers exposed to formaldehyde will be described.

II. CONCEPTUAL BASIS FOR GENETIC MONITORING

Exposure to genetically toxic agents initiates a process which is diagrammed in Figure 1. Cells receiving a significant dose of the agent will experience damage to DNA. This damage

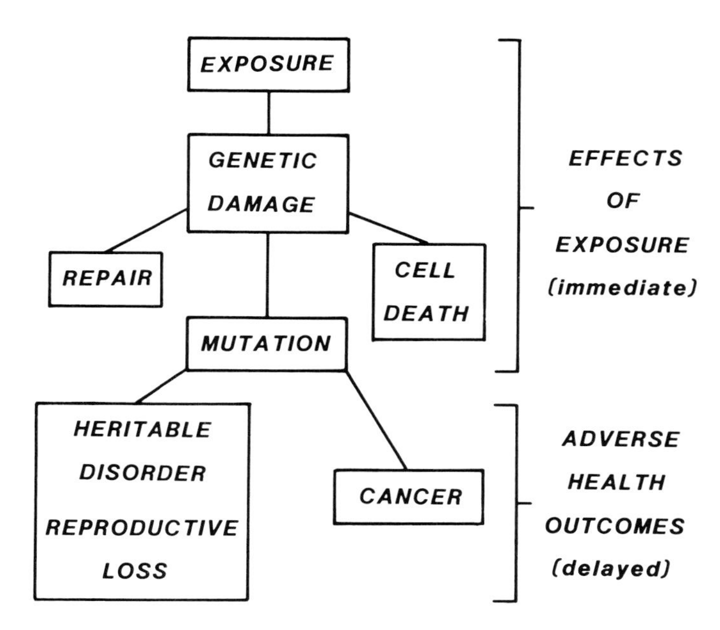

FIGURE 1. Biological consequences of exposure to mutagenic agents. Exposure produces DNA damage which is resolved by cell death, repair, or mutation. These effects are immediately observable and form the basis for short-term genetic monitoring tests. Mutation contributes to the process leading ultimately to cancer. Also mutation in germ cells or embryonic cells can produce reproductive effects. These adverse health outcomes become apparent only after a lag period. As a result they are not as easily detected as the immediate effects. Furthermore, it is more difficult to make an association between the health effects and an exposure than between the immediate effects and exposure.

will be resolved in one of three ways. The damage may be lethal resulting in cell death. DNA repair processes may restore the damaged DNA to its original structure resulting in no ultimate alteration in the cell. Alternatively in the course of repair or replication a permanent change in DNA base sequence, or mutation, may occur. The mutagenic event is thought to play a significant role in the process leading ultimately to neoplasia.[1] Recent molecular studies of oncogene sequences have implicated both base substitution mutation, and gene rearrangement in the activation of oncogenes.[2]

Neoplasia appears after a prolonged latency period as an adverse outcome to which the original mutation contributed. In general, this sequential process is understood to occur in man and animals in fundamentally the same manner. Neoplasia is a relatively rare event. In addition, the long latency of human cancer makes the association of cancers with specific environmental exposures difficult to document. However, various manifestations of the initial genetic damage are observed in a large percentage of treated animals. For example, a chemical may produce tumors in a small percentage of treated animals after a lifetime of exposure, however, the same agent used in acute studies may produce chromosome damage in a high percentage of bone marrow cells immediately following exposure.

The rationale behind the use of genetic damage assays as indicators of exposure is that the events observed are high frequency initial events in a process which ultimately produces neoplastic changes in a small subset of the affected cells. The observation of initial genetic damage in carcinogen-treated animals is the rationale for the use of these tests. In addition, accumulating experience indicates that genetic damage has also been observed in human

Table 1
TESTS AVAILABLE FOR MONITORING HUMAN POPULATIONS FOR GENETIC DAMAGE

Test	Cell or fluid	Biological level
Somatic cell effects		
Structural chromosome aberration	Lymphocyte	Chromosomal
Sister chromatid exchange	Lymphocyte	Chromosomal
6-Thioguanine resistance	Lymphocyte	Gene mutation
Body fluid analysis	Urine	Gene mutation
Macromolecule alkylation	Erythrocyte (hemoglobin) Lymphocyte (DNA)	Molecular-adduct formation
Alkaline filter elution	Lymphocyte	Molecular (DNA single strand breaks, alkali labile sites, crosslinks)
Germinal cell effects		
Sperm count	Semen	General toxicity, including mutagenicity
Sperm morphology	Sperm	Gene mutation (?)
Y fluorescent body frequency (YFF)	Sperm	Chromosome nondisjunction (?)

populations exposed to known carcinogens.[3] The types of events associated with carcinogen exposure include chromosome damage (structural aberrations and sister chromatid exchange), DNA or protein adduct formation, somatic mutations in peripheral lymphocytes, and elevated frequencies of morphologically abnormal sperm or sperm bearing two Y fluorescent bodies. In addition the presence of mutagens in body fluids, particularly urine, has been associated with several chemical exposures.

These events are simply and rapidly detected in cells and body fluids which can be obtained with a minimum of risk or even inconvenience to the subject. In addition these tests can be performed under properly controlled conditions, with ranges of interindividual and temporal variation consistent with many commonly performed clinical laboratory procedures.[4]

III. MONITORING TESTS IN USE

A number of tests have been developed for use in genetic monitoring. Some have been used rather extensively while others are relatively new and not extensively characterized. Different tests can detect events in somatic or germinal cells. Also, events occurring at the molecular and chromosomal levels can be detected. Some of the events are persistent, allowing cumulative effects over a period of time to be evaluated. However, other events are transitory so that events associated with specific time periods can be studied. Table 1 classifies several of the tests now in use or in development with respect to these characteristics. The following sections briefly describe the features of some of the more established monitoring tests.

A. Cytogenetic Tests

By far the greatest body of human genetic monitoring experience has been with cytogenetic tests. Cytogenetic abnormalities are associated wih cancer in several ways. Several human cancers are associated with specific chromosomal rearrangements or deletions. For example, chronic myeloid leukemia cells contain a terminal deletion in chromosome 22 (Philadelphia chromosome)[5] and a reciprocal translocation between chromosomes 8 and 14 is observed in Burkitt's lymphoma.[6] In three inherited human syndromes predisposing to cancer, cells show elevated levels of chromosomal abnormalities: Bloom's syndrome,[7] ataxia telangiectasia,[8] and Fanconi's anemia.[9] Among the short-term tests for carcinogens, assays for chromosome damage have been remarkably accurate.[10]

Cytogenetic analysis for human monitoring is usually done with peripheral lymphocytes using structural damage (chromosome breaks and exchanges) and sister chromatid exchange as the endpoints. The technical aspects of cytogenetic monitoring have recently been reviewed by Galloway and Tice,[11] and by Bloom.[12] The human peripheral lymphocyte is a particularly attractive cell with which to do cytogenetic analysis because its lifetime in the circulation is several years.[13-16] In addition, it can tolerate certain types of genetic damage for long periods.[17] It is easily and safely obtained and reliably cultured by well-established methods. The systematic classification of structural aberrations[18,19] provides a uniform standard for scoring. The sister chromatids of metaphase chromosomes can be differentially stained permitting the observation of reciprocal exchanges of homologous regions (sister chromatid exchange or SCE).[20]

Both structural aberration and SCE analysis have been performed with populations exposed to a variety of agents (reviewed by Sorsa et al.[3]). Increased rates of chromosome damage have been observed in populations exposed to all the agents classified as known or suspect human carcinogens by the IARC for which human chromosome studies have been done.[21] Structural aberrations have been shown to be a sensitive response to certain exposures. Exposures to gamma radiation at doses below the internationally accepted limit of 5 rem/year have produced increased chromosome aberration rates in a population of nuclear dockyard workers.[22]

Small increases in both structural aberrations and SCEs can be detected in relatively small populations. As Figures 2A and B indicate, a 1% increase over a 2% base rate of structural aberrations could be detected in a population of 48 subjects[23] and an increase of 1 SCE per cell over a background of 5 SCE per cell could be detected in about 35 subjects.[49]

B. Sperm Tests

The analysis of sperm is attractive in genetic monitoring because of the opportunity to observe effects in germ cells. Currently three types of endpoints have been evaluated as indicators of genotoxic exposure: sperm count, morphology, and the frequency of Y bodies. These methods have recently been reviewed by Wyrobek et al.[24]

Reduction in sperm count is a response to many different conditions and is not a specific response to mutagenic agents. Nevertheless it is a worthwhile parameter to ascertain because it is a good indicator of the general quality of the sample being evaluated. Furthermore, in at least the case of dibromochloropropane (DBCP) dramatically reduced sperm count was the presenting effect which indicated a health hazard in occupational exposure.[25] Sperm counts can be quickly and easily determined by standardized methods.[26] Although sperm counts vary, analysis of sperm for morphological abnormalities is based on the observation that head morphology appears to be under strict genetic control in the mouse and that exposure to mutagenic agents leads to changes in morphology.[27] Human sperm can be classified into several morphological types and the distribution of different types determined. Wyrobek et al.[24] have developed a standardized protocol for scoring sperm morphology which can effectively stabilize variation in scoring over time and even among scorers.

The YFF test is based on the observation that the human Y chromosome stains so intensely with quinacrine that it can be seen in interphase cells as a bright spot.[28] About 40 to 50% of sperm are observed to contain a fluorescent spot or Y body (YF). About 1% of sperm contain two Y bodies (YFF).[29] The second Y body has been proposed to be a second Y chromosome,[29] although this point remains to be established. The putative cause of the

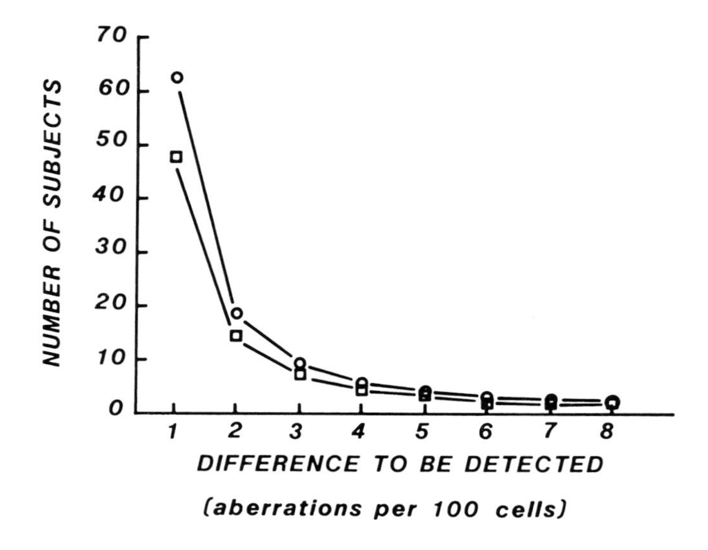

A

NUMBER OF SUBJECTS

80
70
60
50
40
30
20
10
0

1 2 3

DIFFERENCE TO BE DETECTED

(SCE per cell)

B

FIGURE 2. Sample size requirements for cytogenetic monitoring studies. (A) Structural aberration analysis. The number of subjects required to detect indicated increases over a control aberration rate of 2/100 cells is shown when the number of cells scored is 100/subject (O) or 200/subject (□). The error limits used are $\alpha = 0.05$; $\beta = 0.20$. Data are from Reference 23. (B) SCE analysis. The number of subjects required to detect an indicated increase in SCE/cell is shown when the control rate is 5 SCE/cell (O) or 11 SCE/cell (□) and 25 cells/subject are scored. The error limits used are $\alpha = 0.05$ and $\beta = 0.05$. This analysis was made by E. B. Whorton, Jr., using data supplied by R.R. Tice.

appearance of a second Y chromosome is nondisjunction occurring at meiotic anaphase II. Kapp and Jacobsen[30] evaluated the stability of the frequency of the second Y body and its response to exposure to X irradiation and adriamycin. The YFF frequency increased from 1% to 3 to 5% starting about 21 days following the onset of these exposures.

The ability of these sperm tests to detect occupational exposures has been demonstrated in a limited number of cases. Sperm count was reduced in workers exposed to DBCP[25] and lead.[31] DBCP was also associated with increased YFF frequency.[31] Abnormal sperm morphology has been associated with exposure to lead[31] and carbaryl.[33]

Wyrobek et al.[33] have estimated sample size requirements for all three sperm tests. Mean increases over control values of 0.25 standard deviation units could be detected with p = 0.05 and a power of 90% with sample sizes calculated to be 214, 26, and 41 individuals for sperm count, morphology, and YFF, respectively. Except for sperm count, these sample size requirements are similar to those for cytogenetic studies.

C. Body Fluid Analysis

The detection of mutagenic substances in body fluids is an indication of the ingestion, distribution, and excretion of a substance or substances capable of producing genetic damage. Urine samples are tested for the presence of mutagens using an in vitro short-term test. The Ames/Salmonella assay[34] has been the most commonly used although tests for point mutation or cytogenetic effects using mammalian cells or other organisms can be used.[35,36] Similarly, blood or other body fluids could be used as sample sources. Although urine can be tested directly it is common practice to concentrate a nonpolar fraction from the urine using XAD-2 resin extraction.[37] This serves the dual purpose of concentrating mutagens and of removing histidine from the material tested. Low histidine levels interfere with the Ames/Salmonella assay.[38] The original concentration technique[37] is not optimal for concentrating all mutagens. Consequently it is prudent to conduct preliminary tests with urine from treated animals or human urine to which test agents have been added to ensure that the procedure efficiently concentrates the materials of interest in a particular study.

The detection of mutagens in urine is a nonspecific procedure since any agent present will be detected. For situations in which exposures are poorly characterized this is an advantage since the technique is not dependent on the detection of specific agents. If mutagenic activity is found additional urine samples can be subjected to chemical analysis to identify the mutagen and determine its source.[38]

On the other hand, the nonspecific nature of the assay makes it susceptible to interference from other sources of exposure, for example, cigarette smoking.[37,39] Proper experimental design, as described below can adequately control for other sources of variation.

The analysis of urine has detected mutagens in several occupational exposures[3] including oncology nurses and pharmacists involved in preparing and administering cancer chemotherapy drugs as reviewed by Falck.[40]

D. Other Assays

Several other assay systems are in various stages of development at this time which are suitable or potentially useful for monitoring human populations. Most notably, a technique for quantitating specific locus mutation in peripheral lymphocytes is now available.[41] Techniques exploiting new immunological methods and DNA analysis are being used in the development of new assays which are reviewed by Baan et al.[42] in this volume.

IV. INTERPRETATION

In interpreting results of these assays it must be clearly understood that the biological events detected are not adverse health outcomes in themselves. While biological events such as chromosome breakage are associated with exposure to carcinogens, the mechanistic relationship between these events and the induction of neoplasia is not currently understood. In addition all of these events are known to occur in the general population at a finite rate. Recognition of these facts has important implications for the use and interpretation of these tests as well as the way in which results are presented to study participants and sponsors.

The purpose of monitoring tests is to detect biologically significant exposures at an early time point. The lack of a mechanistically understood relationship between the monitoring endpoints and adverse health effects makes results of these studies unsuitable as a basis for a quantitative risk assessment. For example, one cannot extrapolate from a 2% increase in chromosome breakage rate in a population of workers to a precise quantitative increase in risk of some form of cancer.

Furthermore, monitoring tests should be interpreted on the basis of population rather than individual results. This is true for two reasons. The first is, again, the lack of a mechanistic understanding of the relationship between the test endpoint and the adverse health outcome. The second is the range of normal variation among individuals in the general population. There is currently no basis for concluding that an individual in a group under study with the greatest level of genetic damage is at the greatest risk of experiencing an adverse health outcome. Rather the view that should be taken is that an increase in the mean level of damage experienced by a population with a relatively homogenous exposure, as compared to a suitable control group, indicates a significant level of exposure. This exposure implies an increased risk which cannot be precisely quantified. In the absence of more specific data on individual exposure the risk should be considered the same for all individuals in the population regardless of their individual test results.

A corollary is that data resulting from monitoring studies cannot be used to identify individuals who are hypersusceptible to the effects of mutagen or carcinogen exposure. On the basis of monitoring data alone there is no way to differentiate between hypersusceptibility and elevated exposure. Furthermore, elevated responses in short-term tests for genetic damage in individuals have not been shown to imply an increased predisposition to cancer.

For these reasons the results reported to individual participants in monitoring studies should be based on the population rather than their individual findings. Similarly, actions taken in response to monitoring results should be at the population level. Removal of only those individuals with the most abnormal test results from the hazardous environment might well leave other individuals, whose actual risk is just as great, in continued jeopardy. Measures which reduce exposure for the entire population would be more appropriate.

The same characteristics of genetic monitoring endpoints which influence the interpretation of results, as described above, also influence the design of monitoring studies. The comparison of a population under study to a control group is imperative for several reasons. Since several of the endpoints observed are general responses to DNA damaging agents they may be affected by many genetically active substances in the environments of individual subjects. Cigarette smoking, for example, is a common source of exposure to carcinogens which has been shown to increase rates of chromosome damage,[43] SCE,[44] and urinary mutagen excretion.[37] Variability due to exposure to other common mutagens can be controlled by matching participants in exposed groups with control subjects on the basis of sex, age, and lifestyle factors such as smoking or alcohol consumption. Use of medications and other highly individual factors may be difficult to use as matching characteristics but it is important to obtain information on these exposures so that their influence can be determined in interpreting results. By matching individuals in exposed and control groups the general characteristics of the populations can be made similar so that comparison of the populations will focus on the specific exposure differences which are the purpose of the study. The size of the population to be studied is a critical design factor. As with any experiment which seeks to detect an increase over a range of normal variation, consideration must be given to the degree of abnormality which one seeks to detect and the statistical power required to detect it. In practical terms this will influence the size of the population to be studied. Given historical background rates for a specific test and the acceptable limits of false positive (alpha) error and false negative (beta) error the population sizes required to detect specific increases in abnormality can be calculated.[23]

Experimental variability due to laboratory factors must be controlled. Some tests require extensive manipulation of samples obtained from subjects in the laboratory. For example, cytogenetic analysis requires in vitro culture of peripheral lymphocytes. Differences in cell culture methods may alter chromosome aberration rates.[45] Laboratory factors can be controlled by simultaneously obtaining and processing samples from matched control and exposed subjects. Consistent use of the same techniques and materials throughout the study is also an important factor in controlling variation. Because evaluation of many of the endpoints used in monitoring studies requires the subjective judgment of the investigator, scoring should be done on coded samples on a blind basis. In some cases uniformity and objectivity of scoring may be aided by the repeated use of laboratory reference standards.[23] Variation due to both subject characteristics and laboratory factors may be further controlled by sampling the study populations repeatedly at regular intervals. This both increases the sample size and diminishes the impact of temporal variables such as episodic outbreaks of viral infection or sporadic abnormal values of unknown cause.

Not all carcinogens interact with the genetic apparatus in the same manner. One agent may evoke a particular response while another agent evokes a different response. A monitoring study which relies on a single test may fail to detect an exposure to which it lacks sensitivity. By using a battery of tests which detects different types of events in different tissues the probability of detecting an effect should be greatly increased. At the same time a negative finding in a battery of tests can be accepted with greater confidence than if only one test was performed.

In summary, although there are many variables associated with human monitoring tests resulting from individual differences and laboratory techniques appropriate experimental design can control them. As a result statistically valid inferences can be drawn as to the experience of specific exposure groups. Monitoring tests based on indicators of genetic damage or mutagen exposure can be used to detect population exposures to hazardous agents or environments.

V. IMPLEMENTATION OF DESIGN FEATURES: A STUDY OF FORMALDEHYDE EXPOSURE IN PATHOLOGISTS

A study conducted by our laboratory was designed so as to take into account the considerations discussed in the preceding section. The study involved 12 individuals employed on the autopsy service of a major medical center hospital where they were exposed to formaldehyde. In addition 7 other individuals with other responsibilities in the Pathology Department involving formaldehyde exposure were also studied. A battery of tests was performed including environmental measurements to determine formaldehyde levels. Insofar as possible animal tests analogous to the human monitoring studies were performed to determine the potential activity of formaldehyde and to optimize methods. The human tests and their animal analogues are presented in Figure 3.

Each subject was matched with a control subject on the basis of age, sex, customary use of tobacco, alcohol, and recreational drugs. Information collected on caffeine consumption, medications, illnesses, and nonoccupational exposures to potential mutagens was used in evaluating results. Exposed and control subjects were sampled simultaneously to control for laboratory factors and subjects were sampled three times at intervals of 2 to 3 months.

Environmental measurements in the autopsy area indicated that exposures occurred in

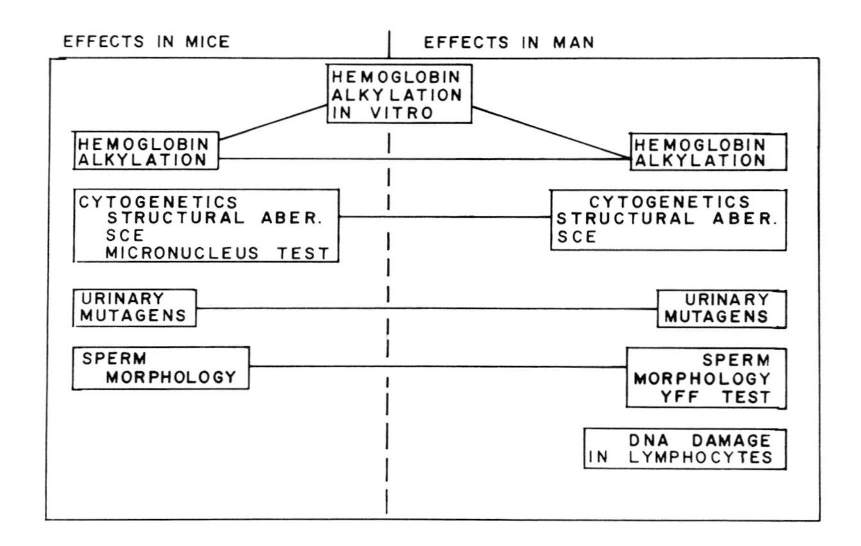

FIGURE 3. Human monitoring tests used in the study of formaldehyde exposure in autopsy service workers and equivalent tests performed in animals. (From Ward, J. B., Jr. et al., in *Short-Term Bioassays in the Analysis of Complex Environmental Mixtures*, Vol. 3, Waters, M. D. et al., Eds., Plenum Press, New York, 1983. With permission.)

transient episodes during specific activities such as sectioning, tissue examination, or autopsy conferences. The time-weighted average (TWA) exposure was estimated to be in the range of 1 ppm.[46]

Cytogenetic evaluation detected no differences between the exposed and control populations for structural aberrations, SCEs, or growth kinetics (Tables 2 and 3). Structural aberration analysis in $B_6C_3F_1$ mice detected a dose-dependent increase in exchange aberrations (centric fusions) (Table 4) between 5 and 20 mg/kg body weight. No increases in micronuclei were observed in mouse bone marrow at doses up to 40 mg/kg body weight. Sperm tests for count, morphology, and YFF detected no increased abnormality in humans and no increases in mouse sperm abnormality were seen. The urine analysis of human subjects detected no increase in mutagenicity although a clustering of subjects with urine extracts toxic to *Salmonella* was seen in the exposed group (Table 5). No mutagenic activity was seen in mouse urine. Studies are still in progress to determine whether adducts to hemoglobin can be detected.

In summary, human exposure to the levels of formaldehyde described did not produce any detectable genetic effects. The animal studies suggest the possibility that much higher levels of exposure could produce chromosome damage. The toxic effects seen with human urine could be associated with formaldehyde exposure although that point remains unresolved.

The population studied in this case was relatively small. In addition, formaldehyde is highly reactive with tissues and rapidly metabolized.[47] It is possible that genetic damage could be produced in cells of the respiratory epithelium without detectable effects ocurring in the more distant cells observed. It would be premature to conclude that human populations exposed to a TWA exposure of 1 ppm are safe from genetic hazards.

VI. FUTURE RESEARCH NEEDS

Several research needs must be met in order to implement human monitoring as an established procedure and to fully realize its potential. The most fundamental problem is to

Table 2
HUMAN EXPOSURE TO FORMALDEHYDE STRUCTURAL CHROMOSOME ABERRATIONS

	High		Low		Total	
Group	Exposed	Control	Exposed	Control	Exposed	Control
n Subjects	12	13	7	7	19	20
n Samples	30	30	20	21	50	51
n Cells	5,591	5,991	3,995	4,080	9,586	10,071
Gaps	38	50	26	32	64	82
Breaks	66	100	58	85	124	185
Exchanges	3	3	3	3	6	6
Other	4	9	2	4	6	13
AB/100[a]	1.31	1.87	1.58	2.25	1.42	2.03
AB/100[b]	1.98	2.70	2.23	3.04	2.09	2.84

[a] Excluding gaps.
[b] Including gaps.

Table 3
FORMALDEHYDE EXPOSURE STUDY — SISTER CHROMATID EXCHANGE

	High		Low		Total	
Group	Exposed	Control	Exposed	Control	Exposed	Control
n Subjects	4	4	6	6	10	10
n Samples	7	8	8	7	15	15
n Cells	201	240	222	210	423	450
% 1st division	69.8	68.0	78.9	71.4	75.2	70.7
% 2nd division	31.0	30.5	20.9	24.8	24.9	27.1
% 3rd division	0.5	1.75	1.5	4.1	0.3	2.9
SCE/cell ± S.D.	6.77	6.02	6.55	6.68	6.64	6.42
	0.96	0.47	1.13	0.67	1.01	0.67

better understand the relationship between the endpoints used in monitoring and the adverse health outcomes which they are intended to predict. Our current lack of understanding is the major reason for the limitations imposed in interpreting test results as discussed earlier. It is also the weakness most often cited as a reason for the reluctance of corporations, unions, or other groups to participate in monitoring studies. The rapid progress now being made in understanding mechanisms of carcinogenesis may help to clarify the significance of some forms of genetic damage. Studies which specifically examine the relationship between genetic damage and tumor induction in animals should be of great interest.

A better understanding of the sensitivity and specificity of monitoring tests in human populations is needed. This requires the participation of many populations with exposures to a variety of agents as well as to different exposure levels. If the specificities of different tests are better defined, batteries of fewer tests tailored to specific situations can be used. This would reduce costs and allow monitoring tests to be conducted on a more routine basis. Similarly, once the sensitivities of specific tests to particular agents is established decisions regarding both test selection and population sample size can be made on a rational basis.

Continued efforts to develop new tests are entirely appropriate. It is by no means clear

Table 4
STRUCTURAL CHROMOSOME ABERRATIONS IN MICE TREATED WITH FORMALDEHYDE

		Formaldehyde				
Treatment/dose	**Water**	**5 mg/kg**	**10 mg/kg**	**20 mg/kg**	**40 mg/kg**	**Cytoxan**
n Animals	10	10	10	10	10	4[a]
n Cells	493	400	500	500	500	63
Aneuploid cells	24	26	66	43	51	9
Gaps	7	3	0	0	1	2
Breaks	2	9	2	2	1	66[b]
Centric fusions	8	13	24	37	16	0
AB/100[c]	2.0	5.5	5.2	7.8	3.4	105
AB/100[d]	3.4	6.25	5.2	7.8	3.6	108

[a] No metaphase cells observed in 6 of 10 animals treated.
[b] Also 3 rings, 1 triradial, 5 severely damaged, 1 moderately damaged.
[c] Excluding gaps.
[d] Including gaps.

Table 5
FORMALDEHYDE EXPOSURE STUDY MUTAGENICITY OF URINE CONCENTRATES

			Result			
Group	**Unit**	**n**	**Neg.**	**Pos.**	**Toxic**	**Quest.**
High exp.	Sample	36	21	2	12	1
	Subject	12	3	1	7	1
High cont.	Sample	36	28	1	5	2
	Subject	13	9	0	3	1
Low exp.	Sample	19	11	0	4	4
	Subject	7	3	0	2	2
Low cont.	Sample	21	18	3	0	0
	Subject	7	6	1	0	0

Exp = exposed; cont. = control; quest. = questionable.

that the tests currently available measure effects most relevant to the neoplastic process. It may be that tests detecting events at the molecular level, such as adduct formation or single strand breakage, or gene rearrangement would prove to be better predictors of neoplasia. In addition, the rapid development of methodologies in molecular genetics and immunology may offer levels of specificity and sensitivity not available today. New tests are also needed which detect effects in specific tissues, particularly those at the site of exposure to direct acting agents. These sites would include the respiratory and digestive tracts and the skin. The lack of sensitivity of current tests to formaldehyde as seen in our study points to the need for such tests in order to clarify the hazards associated with exposure. The emuneration of micronuclei in exfoliated cells described elsewhere[48] is an interesting technique which addresses this problem.

These research needs cannot be met without the willing cooperation of populations of human subjects who are sustaining exposures to potential or known carcinogens in the course of their work or in medical treatment. Studies of sufficient scope and sophistication to meet the objectives of this research will not be possible without the cooperative participation of private industry, organized labor, governmental agencies, and the research community. In order to achieve the level of trust and cooperation needed to conduct meaningful research several social, ethical, and legal issues related to the conduct of studies and the interpretation and use of results need to be addressed and agreed upon. The appropriate use of monitoring data for scientific, regulatory, and legal purposes is an area of particular concern.

The monitoring of human populations for genetic damage has the potential to improve our ability to detect health hazards and prevent or reduce the risk of workplace-exposure-related cancer. In the long run this can reduce costs related to illness. It is possible that the costs of exposure control can also be reduced by better establishing the sources of significant exposure and by understanding the human response to exposure. However, its greatest promise is in the protection of human health through the early detection and prevention of disease processes resulting from mutagenic exposure.

ACKNOWLEDGMENTS

The author wishes to recognize the following individuals who contributed to the formaldehyde exposure study briefly described here: Connie E. Riddle, Rosalind A. Colley, Thomas H. Connor, Virginia Coldiron, Kanokporn Rithidech, Loftan MacMillan, Lina Chang, Michael A. Pereira, and Marvin S. Legator. This study was supported by a grant from the U.S. Environmental Protection Agency (CR 807548).

REFERENCES

1. **Weinstein, I. B.,** Current concepts and controversies in chemical carcinogenesis, *J. Supramolec. Struct. Cell Biochem.,* 17, 99, 1981.
2. **Cooper, G. M.,** Cellular transforming genes, *Science,* 218, 801, 1982.
3. **Sorsa, M., Hemminki, K., and Vainio, H.,** Biologic monitoring of exposure to chemical mutagens in the occupational environment, *Teratogen., Carcinogen., Mutagen.,* 2, 137, 1982.
4. **Kilian, D. J., Moreland, F. M., Benge, M. C., Legator, M. S., and Whorton, E. B., Jr.,** A collaborative study to measure interlaboratory variation with the *in vivo* bone marrow metaphase procedure, in *Handbook of Mutagenicity Test Procedures,* Kilbey, B. J., Legator, M., Nichols, W., and Ramel, C., Eds., Elsevier, Amsterdam, 1977, 243.
5. **Mitelman, F. and Levan, G.,** Clustering of aberrations to specific chromosomes in human neoplasms. III. Incidence and geographic distribution of chromosome aberrations in 856 cases, *Hereditas,* 89, 207, 1978.
6. **Zech, L., Haglund, U., Nilsson, K., and Klein, G.,** Characteristic chromosomal abnormalities in biopsies and lymphoid-cell lines from patients with Burkitt and non-Burkitt lymphomas, *Int. J. Cancer,* 17, 47, 1976.
7. **Chaganti, R. S. K., Schonberg, S., and German, J.,** A many fold increase in sister chromatid exchanges in Bloom's syndrome lymphocytes, *Proc. Natl. Acad. Sci. U.S.A.,* 71, 4508, 1974.
8. **Cohen, M. M., Shaham, M., Dagon, J., Shmueli, E., and Kohn, G.,** Cytogenetic investigations in families with ataxia telangectasia, *Cytogenet. Cell. Genet.,* 15, 338, 1975.
9. **Schroeder, T. M. and German, J.,** Bloom's syndrome and Fanconi's anemia: demonstration of two distinctive patterns of chromosome disruption and rearrangement, *Humangenetik,* 25, 299, 1974.
10. **Radman, M., Jeggo, P., and Wagner, R.,** Chromosomal rearrangement and carcinogenesis, *Mutat. Res.,* 98, 249, 1982.
11. **Galloway, S. M. and Tice, R. R.,** Cytogenetic monitoring of human populations, in *Genotoxic Effects of Airborne Agents,* Tice, R. R., Costa, D. L., and Schaich, K. M., Eds., Plenum Press, New York, 1982, 463.
12. **Bloom, A. D.,** *Guidelines for Studies of Human Populations Exposed to Mutagenic and Reproductive Hazards,* March of Dimes Birth Defects Foundation, White Plains, N.Y., 1981, 4.

13. **Norman, A., Sasaki, M. S., and Ottoman, R. E.,** Elimination of chromosome aberrations from human lymphocytes, *Blood,* 27, 706, 1966.
14. **Buckton, K. E., Smith, P. G., and Court-Brown, W. M.,** The estimation of lymphocyte lifespan from studies on males treated with x-rays for ankelosing spondylitis, in *Human Radiation Cytogenetics,* Evans, H. J., Court-Brown, W., and McLean, A. S., Eds., North Holland, Amsterdam, 1977, 106.
15. **Dolphin, G. W., Lloyd, D. C., and Purrott, R. J.,** Chromosome aberration analysis as a dosimetric technique in radiological protection, *Health Phys.,* 25, 7, 1973.
16. **Bloom, A. D., Neriishi, S., Kamada, N., Iseki, T., and Keehn, R. J.,** Cytogenetic investigation of survivors of the atomic bombings of Hiroshima and Nagasaki, *Lancet,* 2, 672, 1966.
17. **Nowell, P. C.,** Unstable chromosome changes in tuberculin-stimulated leukocyte cultures from irradiated patients. Evidence for immunologically committed, long-lived lymphocytes in human blood, *Blood,* 26, 798, 1965.
18. **Evans, H. J.,** Chromosome aberrations induced by ionizing radiations, *Int. Rev. Cytol.,* 13, 221, 1962.
19. **Savage, J. R. K.,** Classification and relationships of induced chromosomal structural changes, *J. Med. Genet.,* 12, 103, 1975.
20. **Perry, P. and Wolff, S.,** New giemsa method for the differential staining of sister chromatids, *Nature (London),* 251, 156, 1974.
21. **Ward, J. B., Jr., Legator, M. S., Pereira, M. A., and Chang, L. W.,** Evaluation in man and animals of tests for the detection of population exposures to genotoxic chemicals, in *Short-Term Bioassays in the Analysis of Complex Environmental Mixtures,* Vol. 3, Waters, M. D., Sandhu, S. S., Lewtas, J., Claxton, L., Chernoff, N., and Nesnow, S., Eds., Plenum Press, New York, 1983, 461.
22. **Evans, H. J., Buckton, K. E., Hamilton, G. E., and Carothers, A.,** Radiation-induced chromosome aberrations in nuclear-dockyard workers, *Nature (London),* 277, 531, 1979.
23. **Whorton, E. B., Jr., Bee, D. B., and Kilian, D. J.,** Variations in the proportion of abnormal cells and required sample sizes for human cytogenetic studies, *Mutat. Res.,* 64, 79, 1979.
24. **Wyrobek, A. J., Gordon, L. A., Watchmaker, G., and Moore, D. H.,** II, Human sperm morphology testing: description of a reliable method and its statistical power, in *Indicators of Genotoxic Exposure,* Banbury Report 13, Bridges, B. A., Butterworth, B. E., and Weinstein, I. B., Eds., Cold Spring Harbor Laboratory, Cold Spring Harbor, N.Y., 1982, 527.
25. **Legator, M. S.,** Chronology of studies regarding toxicity of 1-2-dibromochloropropane, *Ann. N.Y. Acad. Sci.,* 329, 331, 1979.
26. **Bauer, J. D.,** Semen analysis and infertility investigations, pregnancy tests and placental hormones, in *Clinical Laboratory Methods,* 9th ed., C. V. Mosby, St. Louis, 1982, 736.
27. **Wyrobek, A. J. and Bruce, W. R.,** The induction of spermshape abnormalities in mice and humans, in *Chemical Mutagens: Principles and Methods for Their Detection,* Vol. 5, Hollaender, A. and DeSerres, F. J., Eds., Plenum Press, New York, 1978, 257.
28. **Zech, L.,** Investigation of metaphase chromosomes with DNA binding fluorochromes, *Exp. Cell Res.,* 58 (Abstr.), 463, 1969.
29. **Barlow, P. and Vosa, C. G.,** The Y chromosome in human spermatozoa, *Nature (London),* 226, 961, 1970.
30. **Kapp, R. W., Jr. and Jacobson, C. B.,** Analysis of spermatozoa for Y chromosome non-disjunction, *Teratogen, Carcinogen., Mutagen.,* 1, 193, 1980.
31. **Lancranjan, I., Popescu, H. I., Gavanescu, O., Klepsch, I., and Serbanscu, M.,** Reproductive ability of workmen occupationally exposed to lead, *Arch. Environ. Health,* 30, 396, 1975.
32. **Kapp, R. W., Jr., Picciano, D. J., and Jacobsen, C. B.,** Y-chromosomal nondisjunction in dibromochloropropane exposed workmen, *Mutat. Res.,* 64, 47, 1979.
33. **Wyrobek, A. J., Watchmaker, G., Gordon, L., Wong, K., Moore, D., II, and Whorton, D.,** Sperm shape abnormalities in carbaryl-exposed employees, *Environ. Health Perspect.,* 40, 255, 1981.
34. **Durston, W. E. and Ames, B. N.,** A simple method for the detection of mutagens in urine: studies with the carcinogen 2-acetylaminofluorene, *Proc. Natl. Acad. Sci. U.S.A.,* 71, 737, 1974.
35. **Guerrero, P. R., Rounds, D. E., and Hall, T. C.,** Bioassay procedure for the detection of mutagenic metabolites in human urine with the use of sister chromatid exchange analysis, *J. Natl. Cancer Inst.,* 62, 805, 1979.
36. **Amacher, D. E., Turner, G. N., and Ellis, J. H., Jr.,** Detection of mammalian cell mutagens in urine from carcinogen dosed mice, *Mutat. Res.,* 90, 79, 1981.
37. **Yamasaki, E. and Ames, B. N.,** Concentration of mutagens from urine with the non-polar resin XAD-2: cigarette smokers have mutagenic urine, *Proc. Natl. Acad. Sci. U.S.A.,* 74, 3555, 1977.
38. **Connor, T. H.,** Detection of Chemical Mutagens in the Urine of Humans: Studies of Exposure to Formaldehyde and Cigarette Smoke, Dissertation, University of Texas Medical Branch, Galveston, 1982.
39. **Connor, T. H., Ramanujam, V. M. S., Ward, J. B., Jr., and Legator, M. S.,** The identification and characterization of a urinary mutagen resulting from cigarette smoke, *Mutat. Res.,* 113, 161, 1983.

40. **Falck, K.,** Cytostatic drugs as an occupational hazard, unpublished.
41. **Albertini, R. J.,** Studies with T-lymphocytes: an approach to human mutagenicity monitoring, in *Indicators of Genotoxic Exposure,* Banbury, Report 13, Bridges, B. A., Butterworth, B. E., and Weinstein, I. B., Eds., Cold Spring Harbor Laboratories, Cold Spring Harbor, N.Y., 393, 1982.
42. **Baan, R. A., Lohman, P. M. H., Zaalberg, O. B., Schoen, M. A., Fichtinger-Schepman, A. M. J., Schutte, H. H., and Schans, G. P., v.d.,** Future tools in biomonitoring, in *Carcinogens and Mutagens in the Environment,* Volume 4, Stich, H. F., Ed., CRC Press, Boca Raton, Fla., 1985.
43. **Vijayalaxmi and Evans, H. J.,** *In vivo* and *in vitro* effects of cigarette smoke on chromosomal damage and sister chromatid exchange in human peripheral blood lymphocytes, *Mutat. Res.,* 92, 321, 1982.
44. **Carrano, A. V.,** Sister chromatid exchanges as an indicator of human exposure, in *Indicators of Genotoxic Exposure,* Banbury Report 13, Bridges, B. A., Butterworth, B. E., and Weinstein, I. B., Eds., Cold Spring Harbor Laboratories, Cold Spring Harbor, N.Y., 1982, 307.
45. **Dabney, B. J.,** The role of human genetic monitoring in the workplace, *J. Occup. Med.,* 23, 626, 1981.
46. **Coldiron, V. R., Ward, J. B., Jr., Trieff, N. M., Janssen, H. E., Jr., and Smith, J. H.,** Occupational exposure to formaldehyde in a medical autopsy service, *J. Occup. Med.,* 25, 544, 1983.
47. International Agency for Research on Cancer, *Formaldehyde IARC Monographs on the Evaluation of the Carcinogenic Risk of Chemicals to Humans,* Vol. 23, IARC, Lyon, 1982, 345.
48. **Stich, H. F. and Rosin, M. P.,** Micronuclei in exfoliated human cells as an internal dosimeter for exposures to carcinogens, in *Carcinogens and Mutagens in the Environment,* Vol. 2, *Naturally Occurring Compounds: Endogenous Formation and Modulation,* Stich, H. F., Ed., CRC Press, Boca Raton, Fla., 1984.
49. **Whorton, E. B., Jr.,** Personal communication.

Chapter 11

FUTURE TOOLS IN BIOMONITORING

R. A. Baan, P. H. M. Lohman, O. B. Zaalberg, M. A. Schoen, A. M. J. Fichtinger-Schepman, H. H. Schutte, and G. P. Van der Schans

TABLE OF CONTENTS

I. INTRODUCTION

The almost simultaneous discovery of the experimental and human carcinogenicity of vinyl chloride monomer (VCM)[1,2] alerted scientists and regulatory authorities to the possibility that perhaps more, hitherto unsuspected, chemicals could prove to be carcinogenic. Shortly beforehand, Ames and co-workers[3] had perfected the Salmonella short-term test for mutagenicity. Ames's demonstration that many carcinogens were mutagenic in his test system[4] seemed to suggest a possible solution: test all chemicals for mutagenicity and it may be feasible to detect all or most, as yet unknown, carcinogenic hazards. Successes with the food preservative AF-2 in Japan[5] and the flame retardant Tris in the U.S.,[6,7] demonstrated the effectiveness of the method. Various short-term tests based on the direct and indirect detection of the capacity of a compound to alter the genome structure or DNA code in prokaryotic and eukaryotic cells (genotoxic compounds), were developed in the same period.[8] On the basis of the successes obtained with short-term tests, regulators in the U.S. and Europe wanted to introduce these assays in a legal system designed to prevent the unsuspected introduction of new carcinogenic chemicals.[9,10]

Soon, however, several difficulties became apparent. Purchase[11] pointed out that even when mutagenicity tests were 90% accurate in predicting the presence or absence of carcinogenic activity, the *number* of false positives and false negatives depended upon the prevalence of carcinogens in a given sample. This means that the predictability of carcinogenic activity by short-term tests is not much better, or may even be worse, than flipping a coin, if the prevalence of real carcinogens in the sample drops to 10% or lower. The predictive value of short-term tests for genetic effects has recently been reviewed by Committee 1 of ICPEMC[12] (the International Commission for Protection Against Environmental Mutagens and Carcinogens); the many factors preventing a direct and accurate prediction of carcinogenic activity by short-term tests have been reviewed by Committee 2 of ICPEMC.[13] This does not mean, however, that short-term tests have become less important. Ashby,[14] for instance, described with examples how the consideration of nongenetic factors such as pharmacokinetics, metabolism, solubility, or half-life in physiological surroundings can lead to a conclusion about whether the result of a certain short-term test for a given chemical is relevant. Even such an optimal usage of short-term tests, however, cannot provide more than a qualitative verdict on the possibility that the compound tested is a human hazard.

Perhaps the most striking report on the application of short-term tests was produced by Committee 3 of ICPEMC.[15] In this report a first attempt was made to critically review the scientific basis of legal requirements for the results of short-term tests. The report demonstrated that it is virtually impossible to interpret the results of short-term tests on their own on a regulatory basis. Still, the regulatory duties established by law soon confronted authorities with the difficult task of establishing safe levels of exposure to carcinogens like VCM. That this may not be entirely impossible, is indicated by the experience with isoniazid (INH). This antituberculosis drug is a mouse carcinogen for which excellent epidemiological information exists, indicating that it may not be a carcinogenic hazard to man.[16]

In the case of VCM, Anderson et al.[17] in the U.K. and Natarajan[18] in The Netherlands, evaluating the possible risk of human exposure to low concentrations of this chemical, established independently that an increase in chromosome aberrations in the blood lymphocytes of plant workers, which had been observed after exposure to high concentrations of VCM,[19] did not occur at the very low levels achieved in properly protected manufacturing plants. The chromosome aberration test for biomonitoring purposes, however, is expensive, time-consuming and laborious, possibly insensitive, and nonspecific for the chemical to be tested.

Ideal opportunities for scientific investigations may be given by antineoplastic agents, since a rapidly increasing cohort of young people have been cured from the original cancer by treatment with precisely recorded dosages of cytostatic agents.[20]

Due to the generally unknown effects of factors such as absorption and metabolism[21] of the test chemical or individual and species differences in susceptibility to the compound, probably resulting — at least in part — from differences in cellular repair mechanisms,[22] it would be a great step forward if it became possible to determine directly the extent of the actual toxicologically relevant damage, preferably at the target cell level.

For a genotoxic agent the cellular target is DNA. Therefore, in recent years, attention has been devoted to the development of sensitive methods to measure damage at the DNA level, which will allow a quantitative comparison between exposure dose and effective dose of a given genotoxic agent in the exposed organism. Such biological monitoring, based on molecular dosimetry of DNA lesions with the help of biochemical and immunochemical techniques, is currently under investigation.[23] The aim of these studies is to identify and quantify DNA adducts in humans exposed to specific genotoxic chemicals. Moreover, the occurrence and persistence of specific DNA lesions relevant to certain biological endpoints (like cancer) in different organs of one species, and in organs of different species can be measured with the help of these techniques. This will provide additional information for the extrapolation from data obtained in animal short-term and long-term assay systems to human hazard evaluation.

II. DEVELOPMENT OF METHODS TO MONITOR EXPOSURE TO GENOTOXIC AGENTS

Our studies are aimed at establishing methods to detect human exposure to genotoxic substances via analysis of blood samples, which can be obtained readily and on a routine basis. To obtain insight into the relevance of blood monitoring for genetic effects in other organs, these methods will additionally be evaluated in animal experiments; thus complementary research can be done with destructive methods.

A. Biochemical Techniques for Biomonitoring

1. Detection of DNA Strand Breaks After Exposure to Ionizing Radiation or Alkylating Agents

Ionizing radiation has been connected with neoplasms in various organs.[24] Epidemiological studies have revealed that the enhanced incidence of cancer among people who have been exposed to ionizing radiation can be correlated with relatively low radiation doses. In the case of prenatal irradiation, carcinogenic effects of ionizing radiation have even been demonstrated to occur at doses as low as 0.05 Gy.[25] The most intensively studied DNA lesions resulting from exposure to ionizing radiation are strand breaks. In the past decade, very sensitive techniques have been developed to measure single- and double-strand breaks. These methods, viz., alkaline and neutral elution,[26,27] are based on the fact that the rate at which DNA is eluted through a small pore-size filter, is proportional to the amount of strand breaks present in the DNA. To develop these methods for biological monitoring of radiation damage, experiments were initiated recently, in which human lymphocytes were exposed in vitro to low doses of γ-rays, and subsequently analyzed for DNA damage. Routinely, when, e.g., human fibroblasts are studied after irradiation, the cells are prelabelled with [^{3}H]-thymidine prior to analysis of radiation damage by means of alkaline or neutral elution. In the preliminary experiments mentioned here and in future biomonitoring studies, where radiolabeling is not feasible, the DNA in the eluted fractions is measured fluorometrically.[28,29] Recent results indicate that DNA damage in human lymphocytes exposed in vitro to low doses of ionizing radiation (1 Gy), can be measured with high accuracy (Figure 1A).

Alkaline elution has also been applied to monitor in vitro exposure of human lymphocytes

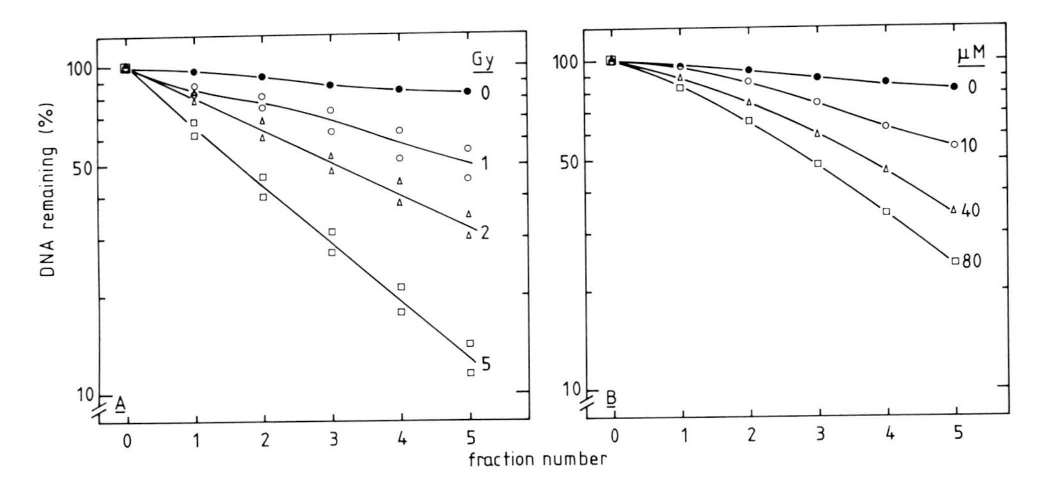

FIGURE 1. Alkaline elution through membrane filters of DNA from human lymphocytes irradiated in vitro with ^{60}Co-γ-rays, in air at 0°C with 0, 1, 2, and 5 Gy (A), or treated with *N*-ethyl-*N*′-nitro-*N*-nitrosoguanidine (ENNG), for 10 min at 37°C with 0, 10, 40, and 80 μ*M* (B). Elution was performed at 20°C at a rate of 0.13 mℓ/min; 5 fractions of 4 mℓ were collected and the amount of DNA remaining on the filter was calculated after quantification of the DNA in each fraction on the basis of the fluorescence of a DNA/Hoechst 33258 complex.

to alkylating agents. These agents can give rise to several types of adducts in DNA (see below), some of which are converted into strand breaks under the conditions of alkaline elution (alkali-labile sites). The results of such an experiment are illustrated in Figure 1B.

A dose-response curve, based on the results of the alkaline elution experiments shown in Figure 1A, is given in Figure 2A. Repair of the radiation damage could also be studied with this technique. After irradiation, the cells were incubated over various time intervals before DNA damage was measured. From the results (depicted in Figure 2B), it would appear that these strand breaks do not represent a reliable parameter for biomonitoring purposes, because over 80% of the observed strand breaks disappear within 30 min of irradiation. The results shown in Figure 1B, obtained after treatment of human lymphocytes with an alkylating agent, could not be quantified because alkaline elution curves in these experiments are not linear. This is probably due to the fact that DNA breaks are generated during the alkaline elution.

2. Separation and Quantification of Radiolabeled DNA Adducts by Means of High-Performance Liquid Chromatography (HPLC)

Another technique applicable for biomonitoring studies is HPLC. This method is used to separate, identify, and quantify DNA adducts formed after exposure of cultured cells or animals to e.g., radiolabeled alkylating agents. These compounds can react — either directly or after metabolic activation — with various sites in DNA. Reaction at sites involved in base pairing (e.g., O^6 in guanine, O^4 in thymine) is thought to directly induce miscoding because it leads to fixation of the anomalous tautomeric structure.[30] Possible harmful effects of alkylation at other DNA sites are thought to originate from induction of error-prone repair processes.[31,32] Alkylating agents react mainly with N7 of guanine, but this lesion appears to be repaired quite rapidly.[33] Alkylation by more electrophilic alkylating agents, like ethylnitrosourea (ENU), of the less reactive O^6 position of guanine may be more critical. Persistence of O^6 alkylated guanine has been correlated with the occurrence of tumors in various rodent organs,[34] but later experiments cast some doubt on a universal relation between specific DNA damage and cancer.[35] Reaction of alkylating agents with the phosphate groups of

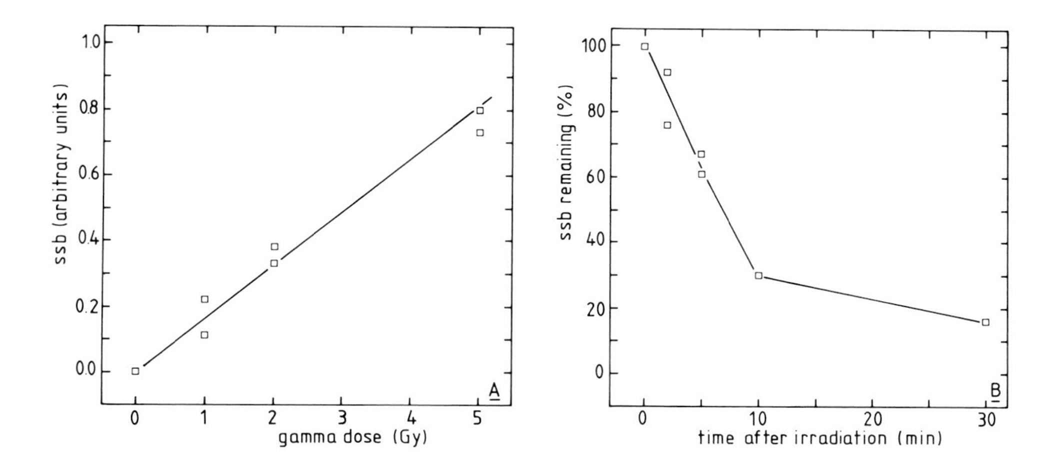

FIGURE 2. The induction and repair of single-strand breaks (ssb) in DNA of human lymphocytes exposed to ^{60}Co-γ-rays. Single-strand breaks were measured by the alkaline elution method. The number of ssb is given in arbitrary units (1 unit corresponds to about 1.6 ssb/10^9 dalton DNA). (A) Dose-response curve. (B) Repair of ssb after a γ-ray dose of 10 Gy and incubation of the cells at 37°C over various time intervals.

nucleic acids has also been demonstrated[36] and the possible importance of the resulting alkylphosphotriesters in mutagenesis and carcinogenesis has been discussed.[37]

Separation of some of the major DNA adducts resulting from exposure of cells or animals to radiolabeled alkylating agents has been achieved by depurination of the isolated DNA under mildly acidic conditions and analysis of the resulting mixture by HPLC on reversed phase and cation exchange columns according to a modified procedure from the literature.[38] An example of the fractionation of ethylated nucleobases on a cation exchange column from the DNA of Chinese hamster cells, treated with [^{3}H]-ENU is shown in Figure 3 and Table 1.

B. Immunochemical Methods for Biomonitoring

The development of immunochemical techniques for sensitive detection of DNA damage has drawn much attention in the past.[39-41] With these methods, a specific DNA lesion can be detected through binding with an antibody directed against that lesion. Other detection methods such as the measurement of DNA repair induced by a genotoxic event are less specific in this respect,[42] while the HPLC method mentioned above requires treatment of cells or animals with highly radioactive genotoxic compounds.

To estimate overall exposure of animals or humans to genotoxic agents by means of immunochemical techniques, antibodies against persistent or long-lived lesions are required. It is essential that these antibodies be highly specific, i.e., recognize a single DNA adduct among a vast excess of unmodified nucleotides (1 in 10^7 or more). At the same time, the antibodies must have a high affinity for the adduct, because the antibody-adduct complex should remain intact under the conditions of the test. The importance of having high-affinity antibodies available to reach the sensitivity required for accurate detection of low levels of DNA damage has been discussed previously.[43]

1. Detection of Acetylaminofluorene (AAF) Adducts in DNA with Specific Antibodies

An enzyme linked immunosorbent assay (ELISA) was applied for sensitive detection of DNA damage.[41,44] To set up the system we used acetylaminofluorene (AAF) as a model genotoxic compound. This aromatic amide is one of the best studied chemical carcinogens.[45] DNA adducts identified after in vivo exposure of rats to AAF comprise *N*-deoxyguanosin-

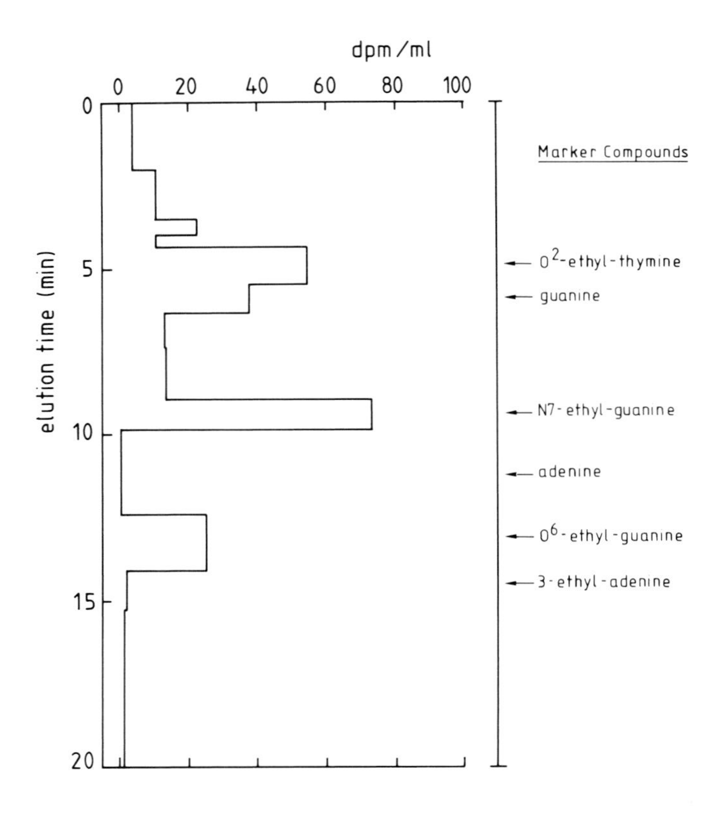

FIGURE 3. Separation of N7-[^{3}H]-ethylguanine and O^6-[^{3}]-ethylguanine from depurinated DNA, isolated from Chinese hamster cells treated with [^{3}H]-ENU, on a HPLC-Rsil SCX (5 μm) cation exchange column. Elution occurred at 1 mℓ/min with 50 m*M* ammonium formiate at pH 4.25 containing 12% methanol; the column temperature was 60°C. The position of co-eluting marker compounds, detected by absorbance at 260 nm, is indicated by arrows.

Table 1
O^6-ETHYLGUANINE AND N7-ETHYLGUANINE IN DNA OF CHINESE HAMSTER CELLS (CHO) TREATED WITH 4 m*M* [^{3}H]-ENU

	Ethylations/nucleotide RSil SCX (HPLC)	
Alkyl adduct	**Exp. 1**	**Exp. 2**
N7-ethylguanine	3.46×10^{-5}	1.89×10^{-5}
O^6-ethylguanine	2.98×10^{-5}	1.23×10^{-5}
Ratio N7/O^6	1.2:1.0	1.5:1.0

8-yl-AAF (dGuo-AAF)[46] and, as a minor adduct, 3-deoxyguanosin-*N*^{2}-yl-AAF.[47] Under certain conditions the C8-adduct is found in the deacetylated form.[48]

We obtained antibodies against the adduct, formed after in vitro incubation of the acetoxy derivative of AAF (AAAF), with deoxyguanosine or deoxyguanosine-5′-monophosphate.

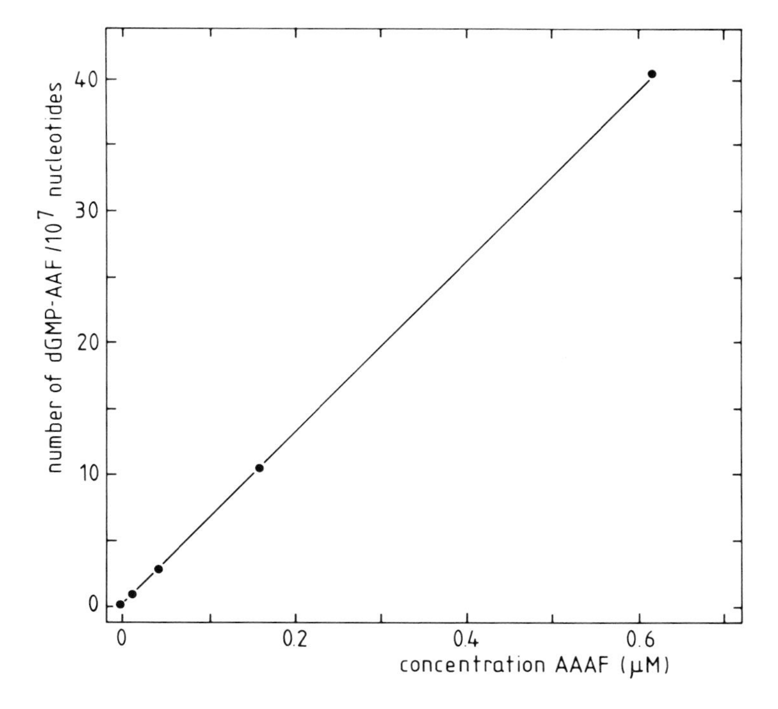

FIGURE 4. Quantitative immunochemical detection of AAF-adducts in DNA, treated in vitro with AAAF. Heat-denatured (10 min, 100°C) salmon sperm DNA was treated for 2 hr at 37°C with various concentrations of AAAF as indicated on the abscissa. After repeated extractions with ether, the samples were dialyzed and incubated in the presence of deoxyribonuclease for 16 hr at 37°C as described previously.[43] An aliquot of each hydrolysate was tested in a competitive ELISA with the rabbit polyclonal anti-dGuo-AAF antiserum. In the same experiment, a calibration curve was constructed by incubation of the antiserum with known amounts of competing antigen (dGMP-AAF).

Under the conditions of our current ELISA, carried out in the competitive mode, the anti-dGuo-AAF antiserum can be diluted about 1.5×10^6-fold. 1 fmol of competing antigen (AAF modified nucleotide, dGMP-AAF) can be detected in 10 μg of DNA (this would correspond to 1 AAF-lesion in about 3.10^7 nucleotides). The result of an experiment in which AAF adducts are measured in a hydrolysate of DNA, treated with various amounts of AAAF, is shown in Figure 4.

2. Use of Monoclonal Antibodies in Biomonitoring

For accurate immunochemical detection of low levels of DNA damage, antibodies are required which have a high specificity and a high affinity for their respective antigen. A conventional polyclonal antiserum contains a variety of antibodies with different specificities and affinities. When minute amounts of DNA adducts are to be detected among a vast excess of unmodified nucleotides, the cross-reactivity of conventional sera with the latter can be a problem.

The discovery[49] that fusion of an antibody-producing cell with a myeloma tumor cell can result in the formation of a continuously growing hybrid cell clone, producing antibodies of uniform specificity and affinity, has focused attention on the preparation of monoclonal antibodies against a large number of different antigens. In recent years, monoclonal antibodies have also been isolated against several types of DNA adducts.[50-52] Especially in the im-

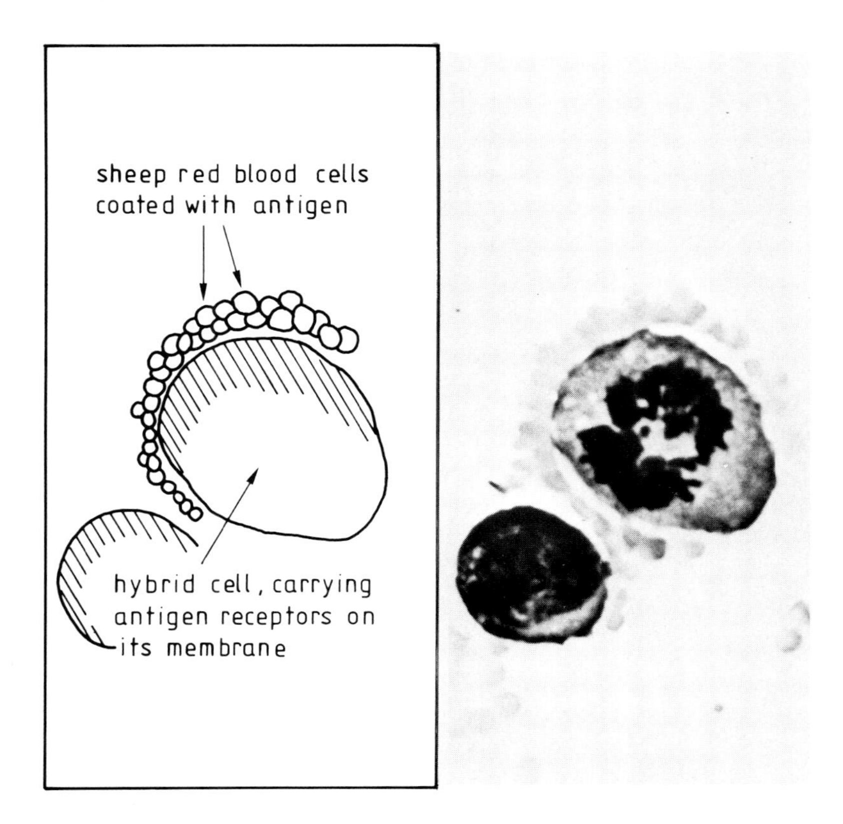

FIGURE 5. Rapid isolation of hybrid cells, producing monoclonal antibodies against dGMP-AAF. Sheep red blood cells (SRBC) were coated with antigen (dGMP–AAF) by incubation in the presence of a water-soluble carbodiimide.[43] The SRBC were washed and added to a fusion mixture of cells containing hybrids producing antibodies against dGMP-AAF (hybrid cells: SRBC = 1:100). The mixture was incubated for 1 hr at 16°C. Rosettes[54] are formed through binding of the antigen on the surface of the SRBC and the antigen receptors on the membrane of the hybrid cells. These rosettes can be easily isolated and grown monoclonally.

munostaining procedures mentioned below, monoclonal antibodies would be a great help, because their specificity and purity ensures a low nonspecific antibody binding.

The hybridoma technique for isolation and selection of monoclonal antibody-producing cell lines involves the growth of many different cell cultures which must be tested regularly for specific antibody production.[53] A more direct method for the selection of hybrid cells that produce the antibodies of interest, is developed in our laboratory. This method is illustrated in Figure 5 for the isolation of monoclonal antibodies against dGuo-AAF adducts.

C. Combined Application of HPLC and Immunochemical Methods

The combined application of HPLC separation techniques and immunochemical methods may enhance the sensitivity of detection of specific DNA adducts because separation of the adducts, by means of HPLC, from the vast excess of unmodified DNA components will decrease the undesired background antibody binding to the latter.

1. Studies on Adducts Formed After In Vitro Treatment of DNA with Cisplatin, an Antineoplastic Agent

The study of the effects of antineoplastic agents is of interest because many of these drugs

Table 2
SPECIFICITY OF ANTIBODIES AGAINST Pt-DNA ADDUCTS

Compound (inhibitor)	IA_{50}[a] (fmol)	Orig. lesion in cisplatin-treated DNA
dGMP	9.10^6	
$Pt(NH_3)_2dGuo/dGMP$	3	
$Pt(NH_3)_2(dGMP)_2$	6	Interstrand crosslink
$Pt(NH_3)_2(pGpG)$	4	Intrastrand crosslink
$Pt(NH_3)_2(pApG)$	3.10^3	Intrastrand crosslink
$Pt(NH_3)_3dGMP$	3.10^3	Monofunctionally bound cisplatin

[a] IA_{50}: the amount required for 50% inhibition in competitive ELISA.

have shown carcinogenic potential in animals and humans.[55] Research on the interactions of such agents with DNA in vitro and in cultured cells may provide insight into the relation between the ensuing DNA damage and the cytostatic activity. In combination with results obtained in animal experiments, these investigations may give information with respect to the synthesis of new analogues with less carcinogenicity and equivalent or higher antitumor activity.

Studies have been initiated to isolate and identify the DNA lesions resulting from treatment of DNA or cultured mammalian cells with cisplatin (*cis*-dichlorodiammineplatinum (II)).[56] Useful information was obtained from experiments in which cell killing and mutation induction by cisplatin were examined in a bacterial system, by studying the induction of nonsense mutations by base-pair substitutions in the *lacI* gene in *Escherichia coli* cells differing in repair capacities.[57] The results indicated that cisplatin causes base-pair substitution mutants only in wild type *E. coli* by the formation of intrastrand crosslinks between two guanines separated in the DNA chain by a third base (GXG; X = A or C). Recently, it was demonstrated by NMR spectroscopy that in vitro reaction between the trinucleotide GCG and cisplatin results in the formation of the expected crosslink product.[58]

In order to separate and identify the various Pt-DNA adducts, salmon sperm DNA was treated in vitro with cisplatin and degraded by enzymatic digestion.[56] The hydrolyzed Pt-containing DNA was analyzed for the presence of various DNA adducts by means of anion exchange chromatography, and analyzed for Pt with flameless atomic absorption spectroscopy (AAS). Among the various Pt-containing (oligo)nucleotides observed, the intrastrand crosslink product of cisplatin on two adjacent guanine residues could be identified by means of NMR and CD spectroscopy. With our present analysis procedure, which uses HPLC anion exchange chromatography (see below), we were also able to identify a Pt-adduct derived from intrastrand crosslinks on the base sequence AG. In addition, there was evidence for the formation of the GXG-Pt adducts mentioned above and for the formation of interstrand crosslinks on two guanines in opposite strands.

Studies with cultured mammalian cells treated with cisplatin in doses allowing for > 10% survival have shown that AAS is not sensitive enough to measure Pt in DNA samples isolated from these cells. Therefore, methods involving immunochemical detection of Pt-DNA adducts are now being developed.[59] We obtained antibodies against Pt-DNA adducts by immunizing rabbits with *cis*-$Pt(NH_3)_2Guo/GMP$, coupled to an immunogenic carrier protein. The specificity of these antibodies was investigated in a competitive ELISA, in which different Pt-nucleotide adducts were added as competing antigens (Table 2). The immunochemical detection of Pt-DNA adducts is carried out in combination with an HPLC separation procedure using an anion exchange column as mentioned above. An example is shown in Figure 6.

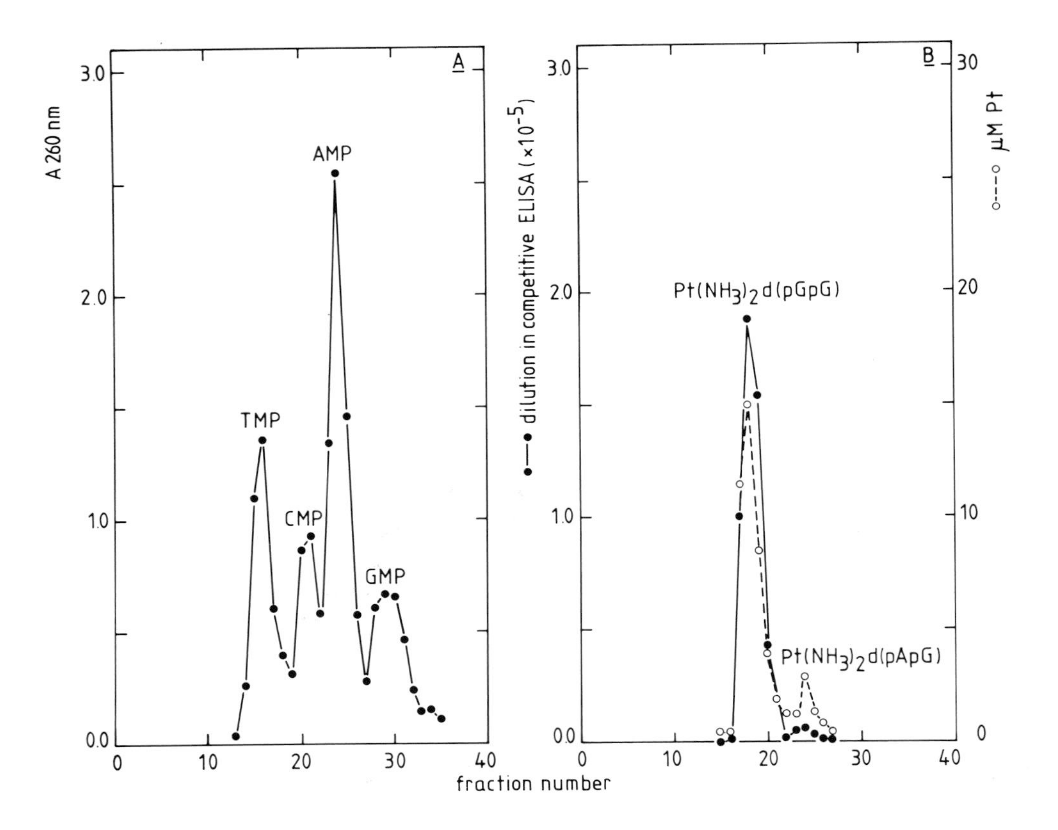

FIGURE 6. HPLC of cisplatin-treated DNA after enzymatic degradation to nucleotides and Pt-containing (oligo)nucleotides, and detection of Pt-DNA adducts in the HPLC fractions by AAS and immunochemical techniques. Salmon sperm DNA was treated in vitro with cisplatin and degraded enzymatically as described previously.[56] The hydrolysate was analyzed on an HPLC Partisil SAX10 anion exchange column, eluted isocratically at 1 mℓ/min with 25 m*M* phosphate, 100 m*M* KCl, pH 6.0, containing 5% methanol. Fractions of 0.5 mℓ were collected. (A) shows the A_{260} profile; (B) shows the Pt-concentration in µ*M* (o — o) as measured with AAS, and also indicates the dilution of the fractions required for 50% inhibition in a competitive ELISA (● — ●).

D. Detection of DNA Damage at the Single-Cell Level

Antibodies directed against specific DNA adducts could be used in studies which relate their presence and persistence in DNA, from cultured cells or animal tissue from various organs, to the eventual biological effect. In such experiments, the occurrence and efficiency of DNA repair processes could be studied by means of antibodies directed against repairable lesions. To estimate overall exposure of animals or humans to genotoxic agents, antibodies against persistent or long-lived lesions are useful. DNA could be isolated from various organs of the animal, whereas nucleated blood cells would be the only readily available source of DNA from humans. In those instances where only limited amounts of tissue are available, methods to detect DNA damage at the level of the single cell are of interest. For such biomonitoring purposes, techniques have been developed in which specific binding of anti-DNA adduct antibodies is measured in the nuclei of fixed cells through detection of fluorescence[60] or radiolabel.

1. Immunofluorescence Microscopy

In the most simple design of this technique, a fluorescent dye is coupled to an anti-DNA adduct antibody in order to localize the DNA adducts in the nuclei of fixed cells in a microscopic preparation. This method can be extended to an indirect immunofluorescence microscopy assay. The cells are first treated with unlabeled antibodies, specific for a certain DNA adduct and, after removal of the unbound antibodies, with a fluorescent anti-immunoglobin antiserum. The second antibody provides for detection of the primary antibody as well as for an amplification of the signal, as several molecules of the second antibody can bind to each primary antibody. Quantification of fluorescence is usually done with microfluorimeter equipment with an image amplifier. With the introduction of image amplifiers which can intensify the brightness up to 10,000 times, computer-based video technology can be applied. The video signal can be digitized and digital processing routines such as noise reduction, background subtraction, and quantitative measurements can be performed on the digitized image.

2. *Autoradiographic Detection of DNA Damage*

A second method for the detection of DNA adducts at the single-cell level is based on quantitative autoradiographic detection of antibodies that recognize specific DNA adducts. This detection occurs through the binding of ^{125}I-labeled protein A from *Staphylococcus aureus*, a protein that specifically binds the Fc-region of immunoglobulins from most mammalian species.[61] Visualization of ^{125}I-protein A binding is achieved by autoradiography. Results of our initial experiments with this method are shown in Figure 7.

III. CONCLUDING REMARKS

Adequate assessment of the risk for human populations from exposure to a genotoxic agent requires extrapolation of the results obtained in animal studies to man and monitoring of the actual exposure of humans to the chemical under study. The induction of genotoxic damage is directly dependent on the relative water/lipid solubility of the compound,[62] and on the dynamic balance between activation and detoxification enzyme systems.[21,63] Once the DNA damage has been inflicted, the presence and the efficiency of repair enzymes determine whether or not the lesion will exert a harmful effect. The fact that the influence of these factors is generally unknown and may even differ widely between species, renders extrapolation of animal data to man very difficult. In animal studies, direct measurement of the actual DNA damage may provide insight into the nature of the attenuating factors mentioned above and facilitate the study of the persistence of the damage and its significance with respect to a certain biological endpoint. These methods can also be applied for biological monitoring in which the exposure of humans to a given chemical is measured on the basis of the presence of specific DNA adducts in, e.g., blood cells. The genetic effects in the human situation, in which the endpoints cannot be measured directly or cannot be related to the initial genotoxic event, can be estimated by making the appropriate comparisons with similar animal data.[64,65] An interesting future approach for extrapolation may be to study both the occurrence of specific DNA damage and the induction of mutations in the hemoglobin gene[66] in human and animal blood cells at the single-cell level.

It should be emphasized that the techniques described in this paper aim at the detection of DNA lesions with a well-defined chemical structure. In those instances, where the structure of the DNA adducts is not known, identification of the various lesions is a prerequisite for the development of a specific biomonitoring assay. The HPLC technique used for the isolation and identification of DNA adducts requires the availability of all possible DNA adducts as marker compounds, whereas for the immunochemical approach, these adducts should be available as antigens for the preparation of antibodies.

Of special interest are the studies directed at the detection of adducts resulting from exposure to Pt-antitumor compounds. These drugs offer the opportunity to directly relate

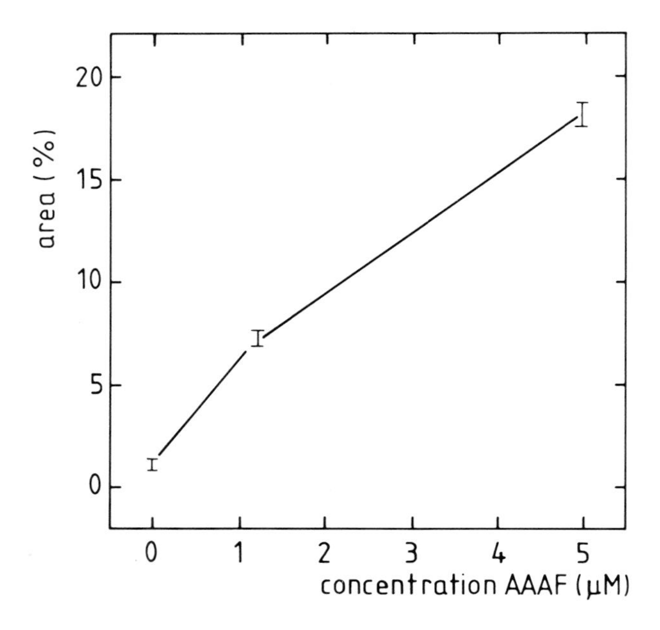

FIGURE 7. Quantitative autoradiographic detection of antibodies that recognize DNA adducts in primary human fibroblasts treated with low doses of acetoxy acetylaminofluorene (AAAF). The subtoxic dose range is between 0 and 1 μ*M*. The preparation of cells or tissue sections for (indirect) immunofluorescence microscopy requires a number of denaturation steps and enzyme treatments in order to make the nuclear DNA accessible for the antibody and to reduce undesired specific staining (e.g., binding to RNA adducts) and nonspecific background fluorescence. The detection of the anti-adduct antibody occurs through the binding of ^{125}I-labeled protein A from *Staphylococcus aureus*.[61] Visualization of ^{125}I-protein A binding is achieved by autoradiography. Quantitation is carried out by counting the developed silver grains over the nuclei of the cells with the use of a microscope and an automatic grain counter (Artek). Corrections are made for the clusters of grains and for the size of the nuclei. The response is expressed as the percentage grain-area over the nuclei.[42]

the in vivo induction of Pt-DNA lesions in human cells to the level of exposure, e.g., by studying the DNA from blood cells of patients treated with accurately known dosages of these drugs.

It is interesting to note that the number of DNA lesions (alkali labile sites, single-strand breaks, and thymine dimers) in mammalian cells resulting from exposure to alkylating agents,[67] ionizing radiation,[68] and ultraviolet light[69] in doses that allow for 37% survival, is in the range between 1 and 1000/10^9 daltons. From the experiments described in this report the detection limits reached so far can be calculated. The results show that 0.25 ssb (after γ-irradiation), 30 ethylguanine adducts (after ENU treatment), 2 AAF-guanine adducts (after AAAF treatment), and 0.1 interstrand crosslinks on opposite guanines (after cisplatin treatment) can be detected per 10^9 daltons DNA. These results demonstrate that, in principle, these techniques may be sensitive enough for biomonitoring purposes in occupational circumstances, especially when the technique for quantitative molecular dosimetry at the single-cell level will be further developed. It should be mentioned in this context that procedures for biomonitoring at the molecular level are not restricted to the detection of DNA lesions. Promising results were obtained, for instance, with the molecular dosimetry of alkyl adducts in blood proteins.[70] However, the disadvantage of the latter system compared to the monitoring of DNA adducts is that protein adducts, contrary to DNA adducts, are nonspecific with regard to the occurrence of specific DNA damage at the target-organ in the human body.

ACKNOWLEDGMENTS

This work was partly sponsored by Shell, the Koningin Wilhelmina Fonds, The Netherlands (projects KWF 81-1 and 83-1), and the European Economic Community (EEC 533-N-(B)).

REFERENCES

1. **Maltoni, C.,** Predictive value of carcinogenesis bioassays, *Ann. N.Y. Acad. Sci.*, 271, 431, 1976.
2. **Waxweiler, R. J., Stringer, W., Wagoner, J. K., Jones, J., Falk, H., and Carter, C.,** Neoplastic risk among workers exposed to vinyl chloride, *Ann. N.Y. Acad. Sci.*, 271, 40, 1976.
3. **Ames, B. N., McCann, J., and Yamasaki, E.,** Methods for detecting carcinogens and mutagens with the Salmonella/mammalian microsome mutagenicity tests, *Mutat. Res.*, 31, 347, 1975.
4. **McCann, J., Choi, E., Yamasaki, E., and Ames, B. N.,** Detection of carcinogens as mutagens in the Salmonella/microsome test: assay of 300 chemicals, *Proc. Natl. Acad. Sci. U.S.A.*, 72, 5135, 1975.
5. **Sugimura, T.,** Mutagens, carcinogens and tumor promotors in our daily food, *Cancer*, 49, 1970, 1982.
6. **Blum, A. and Ames, B. N.,** Flame-retardant additives as possible cancer hazards, *Science*, 195, 17, 1977.
7. Bioassay of tris(2,3-dibromopropyl)phosphate for possible carcinogenicity, Tech. Rep. Ser., 76, Rep. No. 78-1326, National Cancer Institute, Bethesda, Md., 1978.
8. **Hollstein, M., McCann, J., Angelosanto, F. A., and Nichols, W. W.,** Short-term tests for carcinogens and mutagens, *Mutat. Res.*, 65, 133, 1979.
9. Mutagenicity risk assessment: proposed guidelines, *Fed. Regist.*, 45(221), November 13, 1980.
10. Council Directive of 18 September 1979 amending for the 6th time Directive 67-548-EEC on the approximation of the laws, regulations and the administrative provisions relating to the classification, packaging and labelling of dangerous substances (79-831-EEC), in *Off. J. Eur. Communities*, L259, October 15, 1979.
11. **Purchase, I. F. H., Longstaff, E., Ashby, J., Styles, J. A., Anderson, D., Lefevre, P. A., and Westwood, F. R.,** Evaluation of six short term tests for detecting organic chemical carcinogens and recommendations for their use, *Nature (London)*, 264, 624, 1976.
12. Committee 1 ICPEMC, final report: screening strategy for chemicals that are potential germ cell mutagens in mammals, *Mutat. Res.*, 114, 117, 1983.
13. Committee 2 ICPEMC, final report: mutagenesis testing as an approach to carcinogenesis, *Mutat. Res.*, 99, 73, 1982.
14. **Ashby, J.,** The unique role of rodents in the detection of possible human carcinogens and mutagens, *Mutat. Res.*, 115, 177, 1983.
15. Committee 3 ICPEMC, final report: regulatory approaches to the control of environmental mutagens and carcinogens, *Mutat. Res.*, 114, 179, 1983.
16. **Jansen, J. D., Clemmesen, J., and Sundaram, K.,** Isoniazid — an attempt at retrospective prediction, *Mutat. Res.*, 76, 85, 1980.
17. **Anderson, D., Richardson, C. R., Weight, T. M., Purchase, I. F. H., and Adams, W. G. F.,** Chromosomal analysis in vinylchloride exposed workers. Results from analysis 18 and 42 months after an initial sampling, *Mutat. Res.*, 79, 151, 1980.
18. **Natarajan, A. T., Van Buul, P. P. W., and Raposa, T.,** An evaluation of the use of peripheral blood lymphocyte systems for assessing cytological effects induced *in vivo* by chemical mutagens, in *Mutagen-Induced Chromosome Damage in Man*, Evans, H. J. and Lloyd, D. C., Eds., Edinburgh University Press, Edinburgh, Scotland, 1978, 268.
19. **Funes-Cravioto, F., Lambert, B., Lindsten, J., Ehrenberg, L., Natarajan, A. T., and Osterman-Golkar, S.,** Chromosome aberrations in workers exposed to vinylchloride, *Lancet*, 1, 459, 1975.

20. **Adamson, R. H. and Sieber, S. M.,** Antineoplastic agents as potential carcinogens, in *Origins of Human Cancer*, Hiatt, H. H., Watson, J. D., and Winsten, J. A., Eds., Cold Spring Harbor Laboratory, Cold Spring Harbor, N.Y., 1977, 429.
21. **Wright, A. S.,** The role of metabolism in chemical mutagenesis and carcinogenesis, *Mutat. Res.*, 75, 215, 1980.
22. **Rajewsky, M. F.,** Possible determinants for the differential susceptibility of mammalian cells and tissues to chemical carcinogens, in *Quantitative Aspects of Risk Assessment in Chemical Carcinogenesis*, Clemmesen, J., Conning, D. M., Henschler, D., and Oesch, F., Eds., Springer Verlag, Basel, 1980, 229.
23. Development and possible use of immunological techniques to detect individual exposure to carcinogens: IARC, IPCS working group report, *Cancer Res.*, 42, 5236, 1982.
24. **Furth, J. and Lorenz, E.,** Carcinogenesis by ionizing radiations, in *Radiation Biology*, Vol. 1, Hollaender, E., Ed., McGraw-Hill, New York, 1954, 1145.
25. **MacMahon, B.,** Prenatal X-ray exposure and childhood cancer, *J. Natl. Cancer Inst.*, 28, 1173, 1962.
26. **Kohn, K. W. and Ewig, R. A. G.,** Alkaline elution analysis, a new approach to the study of DNA single-strand interruptions in cells, *Cancer Res.*, 33, 1849, 1973.
27. **Fornace, A. J. and Little, J. B.,** Normal repair of DNA single-strand breaks in patients with ataxia-telangiectasia, *Biochim. Biophys. Acta*, 607, 432, 1980.
28. **Parodi, S., Taningher, M., Santi, L., Cavanna, M., Sciaba, L., Maura, A., and Brambilla, G.,** A practical procedure for testing DNA damage *in vivo*, proposed for a pre-screening of chemical carcinogens, *Mutat. Res.*, 54, 39, 1978.
29. **Stout, D. L. and Becker, F. F.,** Fluorometric quantitation of single-stranded DNA: a method applicable to the technique of alkaline elution, *Anal. Biochem.*, 127, 302, 1982.
30. **Loveless, A.,** Possible relevance of O^6-alkylation of deoxyguanosine to mutagenicity of nitrosamines and nitrosamides, *Nature (London)*, 223, 206, 1969.
31. **Radman, M., Caillet-Fauquet, P., Defais, M., and Villani, G.,** Molecular mechanism of induced mutations and an *in vitro* biochemical assay for mutagenesis, in *Screening Tests in Chemical Carcinogenesis*, IARC Sci. Publ. No. 12, Montesano, R., Bartsch, H., and Tomatis, L., Eds., International Agency for Research on Cancer, Lyon, 1976, 537.
32. **Devoret, R., Goze, A., Moulé, Y., and Sarasin, A.,** Lysogenic induction and induced phage reactivation by aflatoxin B1 metabolites, in *Colloque Cancérogénèse Chimique*, Daudel, R., Ed., Centre National de la Recherche Scientifique, Paris, 1976, 283.
33. **Strauss, B. S.,** Repair of DNA adducts produced by alkylation, in *Aging, Carcinogenesis and Radiation Biology*, Smith, K. C., Ed., Plenum Press, New York, 1976, 287.
34. **Goth, R. and Rajewsky, M. F.,** Persistence of O^6-ethylguanine in rat brain DNA, *Proc. Natl. Acad. Sci. U.S.A.*, 71, 639, 1974.
35. **Baur, H. and Neumann, H.-G.,** Correlation of nucleic acid binding by metabolites of *trans*-4-aminostilbene derivatives with tissue specific acute toxicity and carcinogenicity in rats, *Carcinogenesis*, 1, 877, 1980.
36. **Bannon, P. and Verly, W.,** Alkylation of phosphate and stability of phosphate triesters in DNA, *Eur. J. Biochem.*, 31, 103, 1972.
37. **Singer, B.,** The chemical effects of nucleic acid alkylation and their relation to mutagenesis and carcinogenesis, in *Progress in Nucleic Acid Research and Molecular Biology*, Vol. 15, Cohn, W., Ed., Academic Press, New York, 1975, 219.
38. **Beranek, D. T., Weis, C. C., and Swenson, D. H.,** A comprehensive quantitative analysis of methylated and ethylated DNA using high pressure liquid chromatography, *Carcinogenesis*, 1, 595, 1980.
39. **Poirier, M. C.,** Guest editorial: antibodies to carcinogen DNA adducts, *J. Natl. Cancer Inst.*, 67(3), 515, 1981.
40. **Müller, R. and Rajewsky, M. F.,** Review: antibodies specific for DNA components structurally modified by chemical carcinogens, *J. Cancer Res. Clin. Oncol.*, 102, 99, 1981.
41. **Harris, C. C., Yolken, R. H., and Hsu, I.-C.,** Enzymate immunoassays: applications in cancer research, in *Methods in Cancer Research*, Vol. 20, Busch, H. and Yeoman, L. C., Eds., Academic Press, New York, 1982, 213.
42. **Lonati-Galligani, M., Lohman, P. H. M., and Berends, F.,** The validity of the autoradiographic method to detect DNA repair synthesis in rat hepatocytes in primary culture, *Mutat. Res.*, 113, 145, 1983.
43. **Baan, R. A., Schoen, M. A., Zaalberg, O. B., and Lohman, P. H. M.,** The detection of DNA damages by immunological techniques, in *Mutagens in Our Environment*, Sorsa, M. and Vainio, H., Eds., Alan R. Liss, New York, 1982, 111.
44. **Van der Laken, C. J., Hagenaars, A. M., Hermsen, G., Kriek, E., Kuipers, A. J., Nagel, J., Scherer, E., and Welling, M.,** Measurement of O^6-ethyl-deoxyguanosine and N-(deoxyguanosin-8-yl)-N-acetyl-2-aminofluorene in DNA by high-sensitive enzyme immunoassays, *Carcinogenesis*, 3, 569, 1982.

45. **Kriek, E. and Westra, J. G.,** Metabolic activation of aromatic amines and amides and interactions with nucleic acids, in *Chemical Carcinogens and DNA*, Vol. 2, Grover, Ph. L., Ed., CRC Press, Boca Raton, Fla., 1979, 1.
46. **Kriek, E.,** Persistent binding of a new reaction product of the carcinogen N-hydroxy-N-2-acetylaminofluorene with guanine in rat liver DNA *in vivo*, *Cancer Res.*, 32, 2042, 1972.
47. **Westra, J. G., Kriek, E., and Hittenhausen, H.,** Identification of the persistently bound form of the carcinogen N-acetyl-2-aminofluorene to rat liver DNA *in vivo*, *Chem. Biol. Interact.*, 15, 149, 1976.
48. **Poirier, M. C., Williams, G. M., and Yuspa, S. H.,** Effect of culture conditions, cell type, and species of origin on the distribution of acetylated and deacetylated deoxyguanosine C-8 adducts of N-acetoxy-2-acetylaminofluorene, *Molec. Pharmacol.*, 18, 581, 1980.
49. **Köhler, G. and Milstein, C.,** Continuous cultures of fused cells secreting antibody of predefined specificity, *Nature (London)*, 256, 495, 1975.
50. **Haugen, A., Groopman, J. D., Hsu, I.-C., Goodrich, G. R., Wogan, G. N., and Harris, C. C.,** Monoclonal antibody to aflatoxin B1-modified DNA detected by enzyme immunoassay, *Proc. Natl. Acad. Sci. U.S.A.*, 78, 4124, 1981.
51. **Strickland, P. T. and Boyle, J. M.,** Characterisation of two monoclonal antibodies specific for dimerised and non-dimerised adjacent thymidines in single stranded DNA, *Photochem. Photobiol.*, 34, 595, 1981.
52. **Rajewsky, M. F., Müller, R., Adamkiewicz, J., and Drosdziok, W.,** Immunological detection and quantification of DNA components structurally modified by alkylating carcinogens (ethylnitrosourea), in *Carcinogenesis: Fundamental Mechanisms and Environmental Effects*, Pullman, B., Ts'o, P. O. P., and Gelboin, H., Eds., D. Reidel, Dordrecht, 1980, 207.
53. **Oi, V. T. and Herzenberg, L. A.,** Immunoglobulin-producing hybrid cell lines, in *Selected Methods in Cellular Immunology*, Mishell, B. B. and Shiigi, S. M., Eds., W. H. Freeman & Co., San Francisco, 1980, 351.
54. **Zaalberg, O. B., Van der Meul, V. A., and Van Twisk, M. J.,** Antibody production by isolated spleen cells; a study of the cluster and the plaque techniques, *J. Immunol.*, 100, 451, 1968.
55. **Weisburger, J. H., Griswold, D. P., Prejean, J. D., Casey, A. E., Wood, H. B., and Weisburger, E. K.,** The carcinogenic properties of some of the principal drugs used in clinical cancer chemotherapy. Recent results, *Cancer Res.*, 52, 1, 1975.
56. **Fichtinger-Schepman, A. M. J., Lohman, P. H. M., and Reedijk, J.,** Detection and quantification of adducts formed upon interaction of diamminedichloroplatinum(II) with DNA, by anion-exchange chromatography after enzymatic degradation, *Nucleic Acids Res.*, 10, 5345, 1982.
57. **Brouwer, J., Van de Putte, P., Fichtinger-Schepman, A. M. J., and Reedijk, J.,** Base-pair substitution hotspots in GAG and GCG nucleotide sequences in *Escherichia coli* K-12 induced by *cis*-diamminedichloroplatinum (II), *Proc. Natl. Acad. Sci. U.S.A.*, 78, 7010, 1981.
58. **Marcelis, A. T. M., Den Hartog, J. H. J., and Reedijk, J.,** Intrastrand cross-linking of the guanines of the deoxytrinucleotide d(G-C-G) via *cis*-$Pt(NH_3)_2Cl_2$, *J. Am. Chem. Soc.*, 104, 2664, 1982.
59. **Poirier, M. C., Lippard, S. J., Zwelling, L. A., Ushay, H. M., Kerrigan, D. C., Thill, C. R., Santella, M., Grunberger, D., and Yuspa, S. H.,** Antibodies against *cis*-diamminedichloroplatinum (II) –modified DNA are specific for *cis*–diamminedichloroplatinum (II)– DNA adducts formed *in vivo* and *in vitro*, *Proc. Natl. Acad. Sci.*, U.S.A., 79, 6443, 1982.
60. **Slor, H., Mizuzawa, H., Neihart, N., Kakefuda, T., Day, R. S., III, and Bustin, M.,** Immunochemical visualization of binding of the chemical carcinogen benzo(a)pyrene diol-epoxide I to the genome, *Cancer Res.*, 41, 3111, 1981.
61. **Hjelm, H., Hjelm, K., and Sjöquist, J.,** Protein A from *Staphylococcus aureus*. Its isolation by affinity chromatography and its use as an immunosorbent for isolation of immunoglobulins, *FEBS Lett.*, 28, 73, 1972.
62. **Scribner, N. K., Woodworth, B., Ford, G. P., and Scribner, J. D.,** The influence of molecular size and partition coefficients on the predictability of tumors in mouse skin from mutagenicity in *Salmonella typhimurium*, *Carcinogenesis*, 1, 715, 1980.
63. **Glatt, H. R. and Oesch, F.,** Inactivation of electrophilic metabolites by glutathione transferases and limitation of the system due to subcellular localization, *Arch. Toxicol.*, 39, 87, 1977.
64. **Sobels, F. H.,** Evaluating the mutagenic potential of chemicals. The minimal battery and extrapolation problems, *Arch. Toxicol.*, 46, 21, 1980.
65. **Sobels, F. H. and Delehanty, J.,** The first five years of ICPEMC: the International Commission for Protection against Environmental Mutagens and Carcinogens, in *Environmental Mutagens and Carcinogens*, Sugimura, T., Kondo, S., and Takebe, H., Eds., Alan R. Liss, New York, 1982, 81.
66. **Bigbee, W. L., Branscomb, E. W., Weintraub, H. B., Papayannopoulou, Th., and Stamatoyannopoulos, G.,** Cell sorter immunofluorescence detection of human erythrocytes labeled in suspension with antibodies specific for hemoglobins S and C, *J. Immunol. Methods*, 45, 117, 1981.

67. **Abbondandolo, A., Dogliotti, E., Lohman, P. H. M., and Berends, F.,** Molecular dosimetry of DNA-damage caused by alkykation. I. Single strand breaks induced by ethylating agents in cultured mammalian cells in relation to survival, *Mutat. Res.*, 92, 361, 1982.
68. **Van der Schans, G. P., Centen, H. B., and Lohman, P. H. M.,** DNA lesions induced by ionizing radiation, in *Progress in Mutation Research*, Vol. 4, Natarajan, A. T., Ed., Elsevier, Amsterdam, 1982, 285.
69. **Zelle, B., Reynolds, R. J., Kottenhagen, M. J., Schuite, A., and Lohman, P. H. M.,** The influence of the wavelength of ultraviolet radiation on survival, mutation induction and DNA repair in irradiated Chinese hamster cells, *Mutat. Res.*, 72, 491, 1980.
70. **Calleman, C. J., Ehrenberg, L., Jansson, B., Osterman-Golkar, S., Segerbäck, D., Svensson, K., and Wachtmeister, C. A.,** Monitoring and risk assessment by means of haemoglobin alkylation in persons occupationally exposed to ethylene oxide, *J. Environ. Pathol. Toxicol.*, 2, 427, 1978.

Chapter 12

MONITORING OF WORKERS EXPOSED TO BENZENE AND CHLOROFORM

Alessandra Forni and Lorenzo Alessio

TABLE OF CONTENTS

I. INTRODUCTION

Benzene and chloroform are two hydrocarbons widely used in industry and widespread in the environment. Both are important raw materials and good solvents. Both exert toxic effects in humans. Benzene is considered to be a human carcinogen,[1] and chloroform is presently advised to be regarded as such on the basis of animal data.[2] These are, in our opinion, the only reasons that justify putting these two substances in a single chapter, and they will therefore be discussed separately.

For practical reasons, in this context, only chronic exposures which are of interest in industry will be taken into consideration.

II. BENZENE

Benzene (C_6H_6), the first aromatic hydrocarbon, is a highly volatile colorless liquid with a specific odor. At present it is mainly produced in the petrochemical industry, and is mainly used in the chemical industry as a raw material for the synthesis of its derivatives (e.g., phenol, styrene, nitrobenzene, maleic anhydride) and in the manufacture of plastics, pesticides, explosives, drugs, cosmetics, dyes, etc. Due to the use of closed equipments and to hygienically controlled work conditions, cases of high exposure in these industries are exceptional, but chronic low-level exposures may occur. As benzene is also a gasoline component, as a gasoline additive (in concentrations from 1 to 5%, occasionally higher in some countries), refinery workers and those engaged in transport and distribution of gasoline are also potentially exposed.

Benzene, in its commercial form, has largely been used in the past as a solvent in the rubber industry, in shoe manufacture, in rotogravure printing, in the extraction of lipids from seeds, nuts, etc., and these uses induced chronic poisoning in workers exposed to inhalation of high concentrations of the compound. Presently the use of benzene and mixtures containing high concentrations of benzene as solvents is prohibited by law in many countries, and benzene has been substituted by its homologues or by other less toxic solvents. Laboratory workers doing hormone analysis and using benzene for extraction are still potentially exposed.

A. Metabolism and Toxicity

The toxicology of benzene has been the subject of two international workshops held in Paris in 1976 and in Vienna in 1980, and many of the data reported here refer to the conclusions of these meetings.[3,4]

In occupational exposures, benzene enters the body mainly by inhalation of vapors, and the risk increases with increasing temperature. Absorption through the skin is of minor importance for workers of the chemical industry, and absorption by ingestion is negligible.

A proportion (25 to 35%) of benzene inhaled is immediately eliminated through exhaled air, while the remainder enters the blood and is distributed to the various organs, from which it is gradually released. Due to its high solubility in lipids, benzene accumulates in the tissues according to their fatty content; fat and bone marrow are the main sites of accumulation in chronic exposure.

Approximately 50% of absorbed benzene is eliminated unchanged in expired air in 3 successive phases, with a half-life respectively of 40 to 90 min, 2 to 3 hr, and 20 to 30 hr.[5,6] The remainder is metabolized, mainly in the liver, and the metabolites are excreted within 24 to 48 hr via the kidney.

The main metabolic transformation is oxidation by microsomal mixed function oxidases to benzene epoxide, a highly reactive substance which can bind to macromolecules such as proteins and nucleic acids, and is presently considered the most likely metabolite responsible for the toxicity of benzene.

Benzene epoxide can be transformed by a nonenzymatic rearrangement into phenol, or can be hydrated by an epoxide hydratase and then reduced to catechol, or can be condensed with glutathione to form mercapturic acid.

The oxidation phase occurs mainly in the liver, and is followed by a conjugation phase of phenol and catechol with sulfate or glucuronic acid, and conjugates are excreted in the urine. Benzene, however, could be oxidized to epoxide directly in bone marrow cells, such as erythroblasts, and this might account for its hematological toxicity.[7]

On an experimental basis it is suggested that benzene metabolism might be stimulated or inhibited by other chemicals,[7] thus reducing or increasing toxic effects. In humans, liver diseases, in which conjugation processes are impaired, can potentiate the toxic action of benzene.

While acute exposure to high concentrations of benzene exerts mainly a narcotic effect and local irritation of skin and mucous membranes, chronic exposure has mainly a depressant effect on bone marrow, resulting in hyporegenerative anemias of various degree, with insidious onset, known since the last century (see Reference 8 for review). From a critical evaluation of the literature, it emerges that hematological signs (mild anemia, leukopenia, thrombocytopenia) can result from chronic inhalation of benzene at concentrations above 50 ppm and severe hemopathies at concentrations above 100 ppm,[9,10] but there is different individual susceptibility to the hematotoxicity of benzene.

Since the first report in 1928,[11] approximately 200 cases of leukemia occurring singly or in outbreaks, attributable to chronic occupational exposure to high concentrations of benzene, have been described (see reviews in References 8, 9, 12 to 14). In the Italian and Turkish series the leukemias were mainly acute myelogenous in type, while in the French series cases of chronic myelocytic and lymphocytic leukemia are reported.[13,15] In many cases leukemia followed a benzene-induced pancytopenia.

The outbreaks of benzene leukemias observed in the provinces of Milan and Pavia, Italy occurred mainly in shoe factories or in rotogravure plants, where benzene was used as a solvent and where the concentrations of the toxic were very high.[13] The banning of benzene as a solvent has greatly reduced the cases of severe benzene poisoning. No new cases of benzene leukemia have been observed by us in the last 10 years.

The leukemogenic action of chronic exposure to benzene at low concentrations (<25 ppm) is still debated.[3,4] A critical review of the numerous epidemiological studies published from 1974 to date concerning the risk of leukemia in rubber workers, in the printing industry, and in the petrochemical industry, seems to indicate that an excess risk of leukemia, when found, was probably related either to higher exposures that occurred in the past or to multiple exposures,[1,3,4,14] but the problem is still open.

Due to the recognized high toxicity of benzene and its leukemogenicity, it is obvious that exposure should be avoided whenever possible. When there is potential exposure it is mandatory to monitor the working conditions and the populations exposed in order (1) to avoid exposures to dangerous levels and (2) to collect data for future epidemiological studies.

B. Control of Exposed Workers

1. Monitoring of the Workplace

A periodical determination of atmospheric concentrations of benzene in the workplace is necessary to detect the most dangerous jobs and the most exposed workers. This last aim is better fulfilled by means of personal samplers.

The determinations of benzene concentrations should be carried out on different samples collected during the whole shift, at intervals to be defined according to the importance and the characteristics of the pollution.

The permissible levels of benzene differ in various countries. In the U.S. (ACGIH) and in many European countries the time-weighted average (TWA) is 10 ppm (30 mg/m^3), in the USSR the MAC is 5 mg/m^3 (1.5 ppm). The NIOSH in 1976 recommended that no worker should be exposed to concentrations of benzene higher than 1 ppm (3.2 mg/m^3).[16] The recommendations of the Paris workshop, confirmed by the Vienna workshop, were that "in no case should the Time Weighted Average exceed 10 ppm for an 8-hour day and a 40-hour week based on a single work day", and that "based on a 10-15 minute sampling period the ceiling value should not exceed 25 ppm".[3,4]

2. Biological Monitoring

In this context the term "biological monitoring" is used to indicate the evaluation of the exposure of a group of workers or of single workers to toxic chemicals present in the work environment, by measuring the agent and/or its metabolites in samples obtained from the exposed individuals (e.g., blood, urine, exhaled air). These tests are indicators of internal dose.

Biological monitoring of exposure to benzene can be performed by determining benzene concentrations in expired air and in blood, or by measuring its metabolites in the urine of groups of workers at risk. However, due to the scant information on the relationships between atmospheric concentrations and blood levels of benzene, the determination of benzene in blood of exposed workers cannot at present be proposed as an exposure test. On the other hand, since metabolites of benzene are mainly excreted as sulfoconjugates, the determination of the ratio between inorganic and organic sulfates in urine has been used in the past as a method for evaluating benzene exposures, but is no longer used due to its insensitivity for exposures below 50 ppm.

a. Determination of Benzene in Expired Air

Knowledge concerning the elimination of benzene by exhalation and the relationships between atmospheric concentrations of benzene and its levels in expired air is mainly due to the investigations of Sherwood.[5] Since elimination of benzene in expired air follows three phases (see Section II.A), the quantity of the toxic exhaled in the first 2 hr is related mainly to the atmospheric concentrations in the working environment in the period immediately preceding cessation of exposure, while the quantity eliminated after a longer period (16 hr) is related to the total dose absorbed during exposure.

In practice, benzene concentrations in samples obtained immediately before a shift reflect exposure during the previous shift, while the levels in breath samples taken at the end of the shift are greatly influenced by the atmospheric concentrations of benzene toward the end of the shift.

The test is specific and sensitive; however, false positive results may be due to benzene present in cigarette smoke, or to alcohol consumption which increases elimination of the solvent through exhalation.

b. Determination of Phenols in Urine

The determination of total phenols (free and conjugate) in urine is currently used for biological monitoring of benzene exposure. Of the various analytical methods available, the gas chromatographic methods after complete hydrolysis of the conjugated phenols are the most specific.[17]

The values of urinary phenols and the atmospheric levels of benzene are well correlated, as results from several studies indicate (see Reference 17 for review), even though several factors such as individual variability in pulmonary ventilation and in the biotransformation of benzene into phenol make the test valid only on a group basis. When performed in a rather homogeneous group of subjects, the test enables the evaluation of the absorption of the toxic and the hygienic conditions of the workplace. Determinations in single subjects are of more limited value, due to the high individual variability in the spontaneous excretion

Table 1
RELATIONSHIP BETWEEN URINARY PHENOL AT THE END OF THE SHIFT AND BENZENE EXPOSURE, VALID ON GROUP BASIS[3]

Urinary phenol mg/ℓ	Benzene exposure	
	ppm × 8 hr	ppm-hr
<10	Probably 0	
>25	Some	
~50	10	80
~100	25	200

of the metabolite. It is useful, however, to know background pre-exposure values of the workers.

Since the half-life of phenol is 4 to 8 hr and excretion is completed within 48 hr from the end of exposure, the time of urine sampling of exposed subjects is also critical. To evaluate the total absorption of benzene during a shift, urine should be collected at the end of the shift. Background values should be determined on samples obtained at the beginning of a shift cycle after some days of no exposure (e.g., after the weekend).

The relationships between urinary phenols on samples obtained immediately after the end of the shift and benzene exposure, proposed as guidelines by the experts of the Paris workshop,[3] are summarized in Table 1; they are similar to those proposed by Lauwerys.[17]

From these data it is evident that for exposures lower than 5 ppm × 8 hr, the test is not sensitive enough to detect absorption of the solvent.

In evaluating the results, the factors which might increase phenol excretion should be taken into consideration. These include mainly occupational exposure to phenol, the use of phenol-containing ointments, and ingestion of drugs containing this compound or its derivatives (e.g., phenylsalicylate).

Since for practical reasons the test is performed on spot samples, many authors suggest that the analytical values should be adjusted for creatinine or corrected for specific gravity, and that samples having specific gravity <1010 or >1030, or creatinine <0.5 or >3.0 g/ℓ must be discarded.

3. Monitoring of Chromosomal Effects

The determination of chromosome aberration rates in cultured lymphocytes is one of the methods presently available to monitor possible genotoxic effects in populations occupationally exposed to physical or chemical agents. Chromosome studies cannot be considered as a test for biological monitoring of exposure, but as a method for detecting a biological effect, and should therefore be considered on the border between biological monitoring and medical surveillance.

Cytogenetic studies in subjects occupationally exposed to benzene have demonstrated significantly increased rates of chromosomal aberrations both in bone marrow cells and in peripheral blood lymphocytes of subjects with hematological signs of chronic benzene poisoning, in lymphocytes of subjects recovered from benzene hemopathy, as well as in lymphocytes of subjects with past heavy exposure to benzene without signs of toxicity; the data available up to 1976 have been reviewed by Forni.[18] In these subjects increased rates of chromosomal aberrations can persist for years and decades after cessation of exposure, as it is known for radiation-induced chromosome changes.[19-21]

The few studies on groups of subjects with exposure between 5 and 25 ppm have given conflicting results.[18,19,22,23] Positive results in these groups[22,23] can be partly explained by the possible persistence of the effect of past heavier exposures, or by concomitant exposures to other clastogenic agents.

As regards the possibilities and limitations of using chromosome studies to monitor subjects exposed to benzene, or to other mutagens-carcinogens, some points are to be considered, and have been often stressed.[3,4,18,24,25] The main limitations are due to the following facts:

1. The unspecificity of the chromosome damage by various agents, the difficulty in assessing all possible exposures, and the persistence of chromosomal effects in lymphocytes for years may create confounding that must be taken into consideration in evaluating the effects of current exposures.
2. Individual differences both in susceptibility to chromosome damage and in capacity to repair damage and/or eliminate damaged cells, as reflected by dispersion of the data in the groups both of exposed and of controls,[19-21] make these studies valid mainly on a group basis.
3. Although no clear-cut dose-response relationships can be demonstrated for chromosome aberrations in benzene-exposed subjects, it is apparent from the data available that low-level exposures, if effective, induce lower numbers of aberrations. Therefore, in order to demonstrate or exclude the possible cytogenetic effect of a certain exposure, it is necessary to largely increase the sample size both of exposed and of controls (or the numbers of metaphases studied), with consequent increase in cost and technical problems.

The significance of increased rates of chromosome aberrations for the health of the subjects is still largely unknown, but it is a general view that they may be correlated with an increased risk of cancer in the group as a whole, not for single individuals.[24,26]

The findings of chromosome damage in single benzene-exposed subjects in a group per se is not to be considered a disease, but might indicate an individual hypersusceptibility. However, if associated with hematological findings (see ''Medical Surveillance''), further investigations to better assess the etiology of the blood disorder are called for.

The determination of sister chromatid exchanges (SCEs) in lymphocytes, another cytogenetic method presently available to detect a genotoxic effect, seems to be a sensitive indicator of current exposure to some mutagens, even though its use in monitoring populations with chronic low-level exposures has still to be validated and the confoundings are even higher than for chromosome aberrations.[26] A single investigation performed in a group of subjects exposed to benzene up to 6 months previously showed a ''decreased'' rate of SCEs compared to controls.[27]

4. Medical Surveillance

A pre-employment control must be carried out in order to detect conditions of increased susceptibility to benzene, such as blood disorders, liver and renal impairment, and chronic alcoholism. Previous occupational exposure to ionizing radiation or other cytotoxic chemicals should also be evaluated.

Periodical medical examination and complete blood count should be performed at least once a year. The results of the blood count should be carefully compared with the previous ones, particularly with pre-employment values. If alterations of the blood count are detected, the worker should be removed from the dangerous job, since it is known that early cessation of exposure is usually followed by return to normal of the blood values.

Pregnant and nursing women should not be exposed to benzene.

C. Conclusions

On the basis of current knowledge, it is probable that the present conditions of exposure to benzene in industry are sufficiently safe, if the currently accepted TWA of 10 ppm is respected, though one could not completely exclude that occasional cases of leukemia might be induced at these low-level exposures.

As far as biological monitoring is concerned, the determination of benzene in expired air seems to be promising, but must still be validated. The determination of phenol in urine, due to its rather low sensitivity, should be evaluated only on a group basis.

Chromosome studies, due to their low specificity, should be carried out on groups of selected individuals, and the results should be correlated to the exposure in order to determine whether a dose-response relationship exists.

To validate whether the presently suggested TWA is really safe, longitudinal studies of groups of subjects without past heavier exposure to benzene should be started, collecting data on exposure, biological effects, and health status of the workers.

III. CHLOROFORM

Chloroform (trichloromethane, $CHCl_3$) is a highly volatile, colorless liquid, mainly produced by chlorination of methane or by hydrochlorination of methanol.

In industry it is mainly used in the synthesis of fluorocarbon for use as a refrigerant or propellant, and in the production of plastics. Moreover it is used as an extractant in the preparation of dyes, drugs, and pesticides; as a solvent for lacquers, resins, adhesives, polishes, and in many laboratory applications; as an additive in cosmetics and drugs; and as an insecticidal fumigant in agriculture. Potential occupational exposures occur in the production of fluorocarbon, artificial silk, plastics, polishes, in the lacquer industry, in the pharmaceutical industry, etc.

The use of chloroform as an anesthetic has been practically discontinued due to its high toxicity for heart, liver, and kidney. Its use as an ingredient of drugs has been banned in the U.S., Canada, Japan, and in various European countries after the demonstration of carcinogenicity in animal experiments.[2,28]

A. Metabolism and Toxicity

Chloroform can enter the body mainly by inhalation or ingestion. In occupational exposures inhalation is the main route of entrance.

The substance is rapidly distributed to the various organs and tends to accumulate in fatty-rich tissues, such as brain and adipose tissue.

From the scanty data on human metabolism obtained in anesthetized patients, inhaled chloroform results to be partly (30 to 40%) eliminated unchanged by exhalation.[29] No data exist on the fate of inhaled chloroform in occupationally exposed workers. In a single study on 8 volunteers, orally administered ^{13}C-chloroform (0.5 g) was recovered unchanged in expired air in different percentages (from 18 to 67%), with higher retention in obese subjects; in 2 subjects approximately 50% of the administered dose was recovered as $^{13}CO_2$.[30]

Acute exposures to high concentration in humans, besides producing narcotic effects, occasionally induced sudden death and fatal liver and kidney damage. In two studies on workers occupationally exposed to chloroform, an increased incidence of liver enlargement was found.[31,32] No data are available on the carcinogenicity of chloroform in humans.[2]

B. Control of Exposed Workers

As far as monitoring of occupational exposure to chloroform is concerned, the periodical determination of atmospheric concentrations in the workplace seems at present to be the method of choice. The ceiling value for chloroform proposed by NIOSH in 1974 was 240 mg/m^3 (50 ppm), but in 1976, in view of potential carcinogenic hazards, NIOSH recommended that occupational exposure to chloroform should not exceed 9.78 mg/m^3 (2 ppm) in air as determined by a 1-hr air sample.[28]

Due to lack of knowledge on metabolism of chloroform in low chronic exposures, it is impossible to make any practical suggestion for biological monitoring of workers.[33]

No chromosome studies exist on workers exposed only to chloroform. In an in vitro study on human lymphocytes, chloroform with metabolic activation did not induce either chromosome breakage or SCEs.[34]

The medical surveillance of workers both in the pre-employment check-up, and in the periodical examination, should be mainly directed at assessing liver, kidney, and heart functions.

IV. GENERAL CONCLUSIONS

The discovery of the high toxicity and the recognized or potential carcinogenicity of benzene and chloroform have required that the exposures be reduced to very low levels of atmospheric concentrations, which can be measured by the extremely sensitive analytical methods presently available. From the data reported it is apparent that there is a need for further research for sensitive and specific biological indicators to monitor workers chronically exposed to low levels of the toxic substances. For chloroform in particular research is still in its infancy.

As stated also in Section II.C, the collection of data on exposures and biological effects on cohorts of subjects are the basis for future epidemiological studies in order (1) to assess whether the present levels are sufficiently safe, and (2) to detect possible minor biological effects.

One of the major problems is the difficulty in following up for a long time sufficiently large groups, without other occupational exposures or other conditions which can interfere and create confounding. On the other hand, with the present industrial and environmental conditions where multiple exposures may occur, it seems important to also collect data in order to evaluate the possible additive or synergistic effects of various agents in the future.

REFERENCES

1. International Agency for Research on Cancer, *Some Industrial Chemicals and Dyestuff*, IARC Monographs on the Evaulation of the Carcinogenic Risk of Chemicals to Humans, Vol. 29, IARC, Lyon, 1982, 93.
2. International Agency for Research on Cancer, *Some Halogenated Hydrocarbons*, IARC Monographs on the Evaluation of the Carcinogenic Risk of Chemicals to Humans, Vol. 20, IARC, Lyon, 1979, 401.
3. **Truhaut, R. and Murray, R.,** International workshop on toxicology of benzene, Paris: 9th-11th November 1976, *Int. Arch. Occup. Environ. Health*, 41, 65, 1978.
4. **Truhaut, R. and Murray, R.,** International seminar on interpretation of data and evaluation of current knowledge of benzene, Vienna, June 10 and 11, 1980, *Int. Arch. Occup. Environ. Health*, 48, 107, 1981.
5. **Sherwood, R. J.,** Benzene: the interpretation of monitoring results, *Ann. Occup. Hyg.*, 15, 409, 1972.
6. **Sato, A., Nakajima, T., Fujiwara, Y., and Hirosawa, K.,** Pharmacokinetics of benzene and toluene, *Int. Arch. Arbeitsmed.*, 33, 169, 1974.
7. **Snyder, R., Lee, E. W., Kocsis, J. J., and Wither, C. M.,** Bone marrow depressant and leukemogenic actions of benzene, *Life Sci.*, 21, 1709, 1977.
8. **Saita, G.,** Benzene induced hypoplastic anaemias and leukaemias, in *Blood Disorders Due to Drugs and Other Agents*, Girdwood, R. H., Ed., Excerpta Medica, Amsterdam, 1973, 127.

9. Deutsche Forschungsgemeinschaft, Working Group "Establishment of MAK-Werte", Considerations bearing on the question of safe concentrations of benzene in the work environment (MAK-Wert), Bold, Boffard, 1974.
10. **van Raalte, H. G. S.,** A critical look at hazards from benzene in the workplace and community air, *Regulatory Toxicol. Pharmacol.*, 2, 67, 1982.
11. **Delore, P. and Borgomano, C.,** Leucémie aiguë au cours de l'intoxication benzénique: sur l'origine toxique de certaines leucémies aiguës et leur relation avec les anémies graves, *J. Med. Lyon*, 9, 227, 1928.
12. **Forni, A. and Vigliani, E. C.,** Chemical leukemogenesis in man, *Ser. Haematol.*, 7, 211, 1974.
13. **Vigliani, E. C. and Forni, A.,** Benzene and leukemia, *Environ. Res.*, 11, 122, 1976.
14. **van Raalte, H. G. S.and Grasso, P.,** Hematological, myelotoxic, clastogenic, carcinogenic, and leukemogenic effects of benzene, *Regulatory Toxicol. Pharmacol.*, 2, 153, 1982.
15. **Aksoy, M., Erdem, S., and Dinçol, G.,** Types of leukemia in chronic benzene poisoning. A study in thirty-four patients, *Acta Haematol.*, 55, 65, 1976.
16. National Institute for Occupational Safety and Health, *Criteria for a Recommended Standard: Occupational Exposure to Benzene*, U.S. Department of Health, Education, and Welfare, Washington, D.C., 1974, revised 1976.
17. **Lauwerys, R.,** Benzene, in *Human Biological Monitoring of Industrial Chemicals*, Alessio, L., Berlin, A., Roi, R., and Boni, M., Eds., Commission of the European Communities, EUR 8476 EN, Luxembourg, 1983, 1.
18. **Forni, A.,** Chromosome changes and benzene exposure. A review, *Rev. Environ. Health*, 3, 5, 1979.
19. **Tough, I. M., Smith, P. G., Court Brown, W. M., and Harnden, D. G.,** Chromosome studies on workers exposed to atmospheric benzene. The possible influence of age, *Eur. J. Cancer*, 6, 49, 1970.
20. **Forni, A., Pacifico, E., and Limonta, A.,** Chromosome studies in workers exposed to benzene or toluene or both, *Arch. Environ. Health*, 22, 373, 1971.
21. **Forni, A. M., Cappellini, A., Pacifico, E., and Vigliani, E. C.,** Chromosome changes and their evolution in subjects with past exposure to benzene, *Arch. Environ. Health*, 23, 385, 1971.
22. **Funes-Cravioto, F., Zapata-Gayon, C., Kolmodin-Hedman, D., Lambert, B., Lindsten, J., Norberg, E., Nordenskjöld, M., Olin, R., and Swensson, Å.,** Chromosome aberrations and sister-chromatid exchange in workers in chemical laboratories and a rotoprinting factory and in children of women laboratory workers, *Lancet*, 2, 322, 1977.
23. **Picciano, D.,** Cytogenetic study of workers exposed to benzene, *Environ. Res.*, 19, 33, 1979.
24. **Bloom, A. D., Ed.,** *Guidelines for Studies of Human Populations Exposed to Mutagenic and Reproductive Hazards, Proc. March of Dimes Birth Defects Foundation Conf.*, March of Dimes Birth Defects Foundation, White Plains, N.Y., 1981.
25. **Evans, H. J.,** Chromosomal mutations in human populations, *Cytogenet. Cell Genet.*, 33, 48, 1982.
26. **Wolff, S.,** Difficulties in assessing the human health effects of mutagenic carcinogens by cytogenetic analyses, *Cytogenet. Cell Genet.*, 33, 7, 1982.
27. **Watanabe, T., Endo, A., Kato, Y., Shima, S., Watanabe, T., and Ikeda, M.,** Cytogenetics and cytokinetics of cultured lymphocytes from benzene-exposed workers, *Int. Arch. Occup. Environ. Health*, 46, 31, 1980.
28. National Institute for Occupational Safety and Health, *Criteria for a Recommended Standard: Occupational Exposure to Chloroform*, U.S. Department of Health, Education, and Welfare, Washington, D.C., 1974, revised 1976.
29. **Lehmann, K. B. and Hasegawa, A.,** Studien über die Absorption chlorierten Kohlenwasserstoffe aus der Luft durch Tier und Mensch, *Arch. Hyg.*, 72, 327, 1910.
30. **Fry, B. J., Taylor, T., and Hathway, D. E.,** Pulmonary elimination of chloroform and its metabolite in man, *Arch. Int. Pharmacodyn. Ther.*, 196, 98, 1972.
31. **Bomski, H., Sobolewska, A., and Strakowski, A.,** Toxische Schädigung der Leber durch Chloroform bei Chemiebetriebswerkern, *Arch. Gewerbepathol.*, 24, 127, 1967.
32. **Gambini, G. and Farina, G.,** Funzionalità epatica in operai esposti all'inalazione di vapori di chloroformio, *Med. Lav.*, 64, 432, 1973.
33. **Monster, A. C. and Zielhuis, R. L.,** Chlorinated Hydrocarbon Solvents, in *Human Biological Monitoring of Industrial Chemicals*, Alessio, L., Berlin, A., Roi, R., and Boni, M., Eds., Commission of the European Communities, EUR 8476 EN, Luxembourg, 1983, 45.
34. **Kirkland, D. J., Smith, K. L., and Van Abbé, N. J.,** Failure of chloroform to induce chromosome damage or sister-chromatid exchanges in cultured human lymphocytes and failure to induce reversion in *Escherichia coli*, *Food Cosmet. Toxicol.*, 19, 651, 1980.

Chapter 13

IN VITRO SYNERGISTIC EFFECTS OF COMBINATIONS BETWEEN CIGARETTE SMOKE AND ARSENIC, CHROMIUM, OR FORMALDEHYDE

Miriam P. Rosin

TABLE OF CONTENTS

I. INTRODUCTION

Any working environment is exceedingly complex, involving the interaction of numerous factors on exposed individuals. Attempts have been made to disentangle this web of interactions by focusing successively on single components of this environment and identifying them as potential carcinogenic hazards. However, difficulties arise when data obtained from such studies are used to determine an individual's relative risk for cancer. At this point, it is necessary to predict how the combination of various factors (carcinogens, cocarcinogens, promoting agents) to which an individual is exposed will affect his risk of developing cancer. This estimate cannot be obtained merely by summing the relative risk posed by each separate factor. For example, epidemiological studies support a synergistic effect (greater than additive) on risk for lung cancer in populations of smokers exposed to asbestos,[1] or to radiation.[2] Unfortunately, there is a paucity of data on interactions of carcinogenic factors: it is not known which factors in the workplace will interact to produce synergistic effects of cancer risk, which conditions lead to such an interaction, or what the underlying mechanism is for combinational effects.

How do we begin to obtain such information on multifactorial effects? Epidemiological studies, which have uncovered multiple risk factors, lack the power to resolve the mechanism by which such effects are produced. The use of experimental animals for studying multiple factors is also restricted. When more chemicals are applied, the number of combinations and permutations becomes so large that the use of rodents is simply not feasible. However, the highly economic, rapid, and quantitative short-term tests for genotoxicity should be ideally suited for an in-depth examination of multiple interactions between hazardous chemicals.

The focus of this chapter is to indicate how one such genotoxicity assay, the induction of chromosomal aberrations in mammalian cells, can be used to obtain quantitative information on the effects of combinations of environmental factors. The interactions which are presented in this chapter include combinations of cigarette smoke and chromium, cigarette smoke and arsenic, and cigarette smoke and formaldehyde. Each of these combinations would be expected to occur among smokers in a variety of occupational groups. In addition, with the exception of formaldehyde, each of the selected agents possesses carcinogenic activity in exposed human populations: cigarette smoking elevates the frequency of cancers of the lung, the oral cavity and esophagus (especially in combination with alcohol consumption), and urinary bladder;[3,4] arsenic and arsenic-containing compounds are implicated in the development of lung and skin cancers among vine growers, tanners, workers in copper and cobalt smelters, gold mines, and pesticide manufacturers;[5] and chromium is associated with cancer of the respiratory tract in men employed in chromate-producing and chrome pigment industries.[6,7] There are as yet no epidemiological studies clearly demonstrating increased cancer incidence among human populations exposed to formaldehyde. However, the fact that formaldehyde has in vitro genotoxicity in a range of assays[8] and is carcinogenic in animal experiments,[9,10] coupled with the wide usage of formaldehyde in both the workplace (woodworkers, carpenters, embalmers, pathologists, and textile workers) and in daily life (commodities such as hair conditioners, cosmetics, pharmaceuticals, deodorants, toothpastes, foam insulation), make studies on this compound extremely relevant.

II. CIGARETTE SMOKING MACHINE

The majority of studies to date on the genotoxic effects of cigarette smoke have employed either tobacco smoke condensates or isolated fractions thereof.[11] This approach has several limitations, the most obvious of which is the extent to which the chemicals in such condensates resemble the active genotoxic agents in fresh smoke. In addition, results are de-

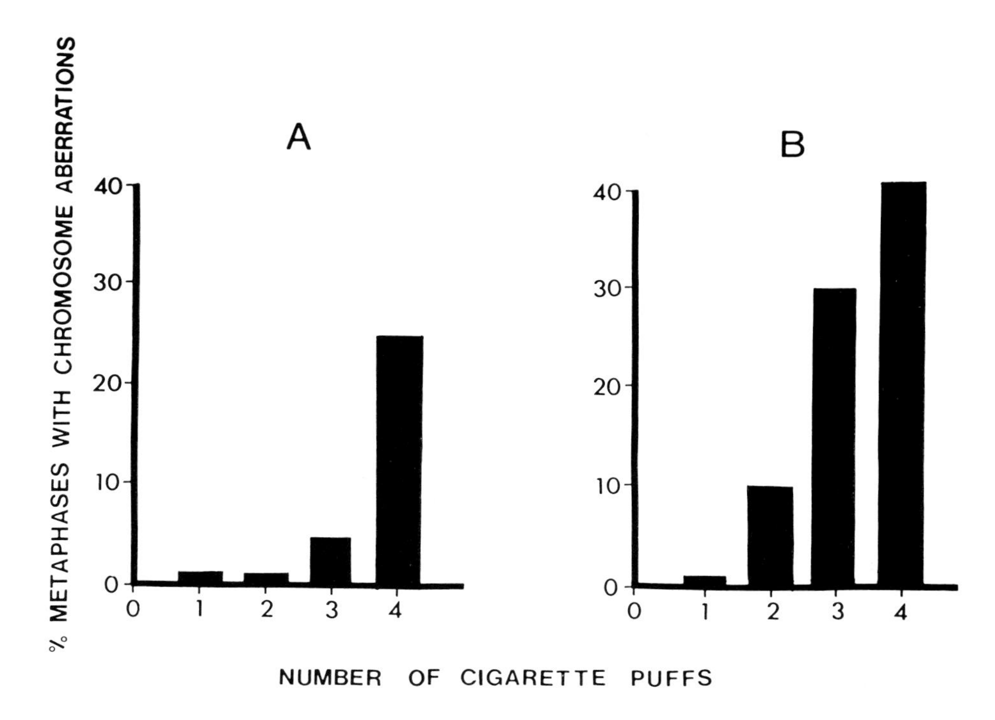

FIGURE 1. Chromosome-damaging capacity of cigarette smoke generated under two experimental conditions by the cigarette smoking machine. The length of the "puff" time is increased from 2 sec in (A) to 4 sec in (B).

pendent upon the particular solvent which is utilized to suspend the condensate into a test solution as well as the choice of an appropriate carcinogen activation system (S9 mix).

In vitro conditions which more closely resemble human exposure patterns can be produced by using a "cigarette smoking machine" to generate cigarette smoke. Mammalian cells are seeded onto glass coverslips and these cultures are exposed directly to cigarette smoke by placing the glass coverslips vertically in a rack in a treatment chamber into which a puff of smoke is blown once a minute. Air entering the chamber is humidified to prevent the cells from drying out. The machine can be adjusted to simulate a variety of smoking patterns (e.g., a fast, nervous smoker or an individual who smokes slowly but inhales deeply) by altering the force and duration of the cigarette puff, the interval between puffs, the volume of fresh air into which the smoke is diluted, and the rate at which the cigarette smoke is cleared from the treatment chamber.

Figure 1 illustrates the chromosome-damaging activity of cigarette smoke on cultures of Chinese hamster ovary (CHO) cells. Two or three puffs of cigarette smoke are sufficient to induce a significant increase in the percentage of metaphases with chromosome aberrations, with both chromatid breaks and chromatid exchanges being induced (Table 1). The genotoxic components are direct acting, with no requirement for a liver microsomal activation system for activity. Alteration of the duration of the puff results in a shift in the dose-response relationship between the number of cigarette puffs employed and the frequency of chromosome aberrations (Figures 1A and 1B)

III. INTERACTION OF CIGARETTE SMOKE AND ARSENIC

The exposure of CHO cells to cigarette smoke generated by the smoking machine can be coupled with a treatment of these cells by environmental agents added to the culture medium.

Table 1
CHROMOSOME ABERRATIONS[a] IN CHO CELLS FOLLOWING SEQUENTIAL EXPOSURE TO SODIUM ARSENITE OR SODIUM ARSENATE AND CIGARETTE SMOKE[b]

		Cigarette smoke in combination with					
		Sodium arsenite ($NaAsO_2$), μM			Sodium arsenate ($NaHAsO_4$), μM		
Cigarette smoke (no. of puffs)	Cigarette smoke alone	20	40	60	200	300	400
0	1.0(0.01)[c]	0.4(0.01)	1.3(0.01)	8.3(0.08)	3.8(0.04)	6.2(0.08)	10.0(0.14)
1	1.0(0.01)	1.3(0.01)	22.0(0.46)	45.0(0.49)	5.0(0.06)	6.0(0.07)	16.5(0.19)
2	4.0(0.04)	22.9(0.33)	40.0(0.76)	MI[d]	20.0(0.31)	25.0(0.31)	37.0(0.44)

[a] The frequency of chromosome aberrations is expressed as (1) the percentage of metaphase plates with one or more chromatid breaks or exchanges, and (2) the average number of chromatid breaks and exchanges per cell (figures in parentheses).
[b] Culture exposed to arsenic solutions for 3 hr followed by exposure to indicated number of cigarette puffs. Colchicine is added to cultures 16 hr after completion of smoking treatment and cultures are harvested 4 hr thereafter.
[c] Untreated control.
[d] MI: Mitotic inhibition; less than 1 metaphase plate among 6000 cells.

This permits one to build model systems for the effect of such combinations at the level of induction of genotoxic activity. Table 1 shows the frequencies of metaphase plates with chromosome aberrations in cultures of CHO cells treated with sodium arsenite (arsenic III) for 3 hr followed by cigarette smoke compared to the effects of either sodium arsenite or cigarette smoke alone. The combination of these two agents results in a synergistic effect on both the percentage of metaphase plates with chromosomal aberrations and on the average number of chromatid breaks and exchanges per cell. This result is particularly apparent when subeffective doses of either agent are combined. For example, exposure to 20 μm sodium arsenite or two puffs of cigarette smoke does not significantly elevate chromosomal aberration frequencies. However, cultures sequentially exposed to both treatments have aberrant chromosomes in 22.9% of the examined metaphase plates. Table 1 also shows the effects of combinations of sodium arsenate (arsenic II) and cigarette smoke. It is important to examine the various oxidative states of arsenic, since the active carcinogenic state of this metal is not yet clearly known, and biological effects of the different oxidation states could be quite different. Sodium arsenate and cigarette smoke also produce a synergistic effect on the induction of chromosome aberrations in treated CHO cells. However, this response requires a tenfold higher concentration of sodium arsenate compared to sodium arsenite.

IV. INTERACTION OF CIGARETTE SMOKE AND CHROMATE

The clastogenic effect of sequential exposure of CHO cells to potassium chromate (chromate VI) and cigarette smoke or to chromic chloride (chromate III) and cigarette smoke is shown in Table 2. Both of these combinations result in a synergistic effect on observed chromosomal aberrations (percent metaphases with chromosomal aberrations and average number of chromatid breaks and exchanges per metaphase plate). However, the doses of chromic chloride which are required to induce clastogenic activity alone or in combination with treatment with cigarette smoke are 100-fold higher than similar doses of potassium chromate.

V. INTERACTION OF CIGARETTE SMOKE AND FORMALDEHYDE

Table 3 presents data showing the effect of sequential exposure of CHO cells to formaldehyde (3 hr) followed by cigarette smoke on chromosomal aberration frequencies. Formaldehyde alone was a potent clastogenic agent. However, in combination with cigarette

Table 2
CHROMOSOME ABERRATIONS[a] IN CHO CELLS FOLLOWING SEQUENTIAL EXPOSURE TO POTASSIUM CHROMATE OR CHROMIC CHLORIDE AND CIGARETTE SMOKE[b]

		Cigarette smoke in combination with					
		Potassium chromate (K_2CrO_4), μM			Chromic chloride ($CrCl_3$), μM		
Cigarette smoke (no. of puffs)	Cigarette smoke alone	10	20	30	800	1000[c]	2000[c]
0	0.5(0.01)	1.0(0.01)	2.0(0.02)	7.0(0.07)	0.4(0.01)	1.0(0.01)	5.0(0.05)
2	2.0(0.02)	4.9(0.06)	20.0(0.23)	38.0(0.43)	7.0(0.09)	9.0(0.10)	12.0(0.13)
3	6.0(0.07)	12.0(0.13)	27.0(0.36)	33.0(0.38)	11.0(0.11)	13.0(0.14)	17.0(0.18)
4	14.2(0.16)	20.0(0.21)	I[d]	I	13.0(0.13)	20.0(0.18)	28.0(0.36)

[a] The frequency of chromosome aberrations is expressed as (1) the percentage of metaphase plates with one or more chromatid breaks or exchanges, and (2) the average number of chromatid breaks and exchanges per cell (figures in parentheses).
[b] 3 hr exposure to chromium solution followed by indicated number of puffs of smoke.
[c] $CrCl_3$ slightly insoluble at this concentration.
[d] MI: Mitotic inhibition; less than 1 metaphase plate among 6000 cells.

Table 3
CHROMOSOME ABERRATIONS IN CHO CELLS FOLLOWING SEQUENTIAL EXPOSURE TO FORMALDEHYDE AND CIGARETTE SMOKE[a]

	% Metaphase plates with at least 1 chromatid break or exchange formaldehyde (μM)			
Cigarette smoke (no. of puffs)	0	20	40	60
0	0.2(0.01)[b]	1.0(0.01)	12.4(0.12)	54.0(1.31)
1	0.4(0.01)	3.8(0.04)	19.0(0.32)	63.0(2.14)
2	10.0(0.04)	28.4(0.36)	58.2(1.38)	77.0(2.09)

[a] Cultures exposed in sealed Leighton tubes for 3 hr to culture media containing formaldehyde followed by exposure to cigarette smoke.
[b] Figures in parentheses are average number of chromatid breaks and exchanges per metaphase.

smoke, the concentration of formaldehyde required to produce a clastogenic effect in the treated cells was reduced from 60 to 20 μm.

VI. DISCUSSION

This chapter presents an ''in vitro'' approach which can be used to assay the effects of combined exposures to environmental agents. The approach uses as an endpoint the ability of such agents to induce chromosomal damage in mammalian cells either alone or in paired combinations. Each of the combinations presented in this chapter (cigarette smoke and arsenic, cigarette smoke and chromium, cigarette smoke and formaldehyde) resulted in a synergistic effect on the frequency of chromosomal aberrations in exposed cells.

The question arises as to whether synergistic effects which are observed in vitro are representative of in vivo situations. Individuals are exposed to a host of agents, including not only carcinogens and cocarcinogens but also various modulating agents (antimutagens, anticarcinogens) which could act to reduce genotoxic damage and cancer risk. On the other hand, in vitro model systems provide the only feasible tool to obtain quantitative data on agent interactions or to perform studies on the mechanism of such interactions. A compromise may be reached by performing initial combination studies with in vitro assays and then extending such studies by using as a second step in vivo tests for genotoxicity on individuals in population groups exposed to such combinations. For example, peripheral lymphocytes from exposed individuals can be examined for the frequency of chromosomal aberrations or sister chromatid exchanges (SCEs). A significant elevation in chromosomal damage in lymphocytes obtained from smokers as compared to nonsmokers has already been observed among smelter workers exposed to arsenic[12] or lead,[13] as well as among employees in electroplating factories exposed to chromic acid.[14] These results are consistent with the in vitro observations presented in this chapter. A third step to such studies would be to look for a correlation between in vivo genotoxicity results showing synergistic effects and cancer statistics. One of the combinations examined in this chapter, cigarette smoking and arsenic, has been reported to produce a significant elevation in the frequency of lung cancer mortality.[15-17]

One method of fine tuning genotoxicity assays so that they may give data more relevant to "cancer risk" assessments is to determine the genotoxic damage in human target cells in which the carcinogen(s) is active. We have recently used such an approach in a study on heavy consumers of alcoholic beverages and tobacco, a combination of agents known to result in a synergistic effect on the induction of oral cancer.[4] A comparison was made of the frequency of micronuclei in buccal mucosa cell samples of heavy smokers, heavy alcohol drinkers, and individuals who were heavy consumers of both items. A synergistic effect was observed on the frequencies of micronuclei in exfoliated cells in individuals consuming both alcohol and tobacco.[18] We are currently extending this approach to other population groups at elevated risk for cancer, with the aim of establishing the assay as an early monitor for carcinogen exposures.

ACKNOWLEDGMENTS

This study was supported by the Associate Committee on Scientific Criteria for Environmental Quality of the National Research Council of Canada. Dr. M. P. Rosin is a staff member of the British Columbia Cancer Foundation of Vancouver, B.C. The author would like to acknowledge the technical assistance of Miss Betty Chow.

REFERENCES

1. **Hammond, E. C., Selikoff, I. J., and Seidman, H.,** Asbestos exposure, cigarette smoking and death rates, *Ann. N.Y. Acad. Sci.*, 330, 473, 1979.
2. **Archer, V. E., Gillan, J. D., and Wagoner, J. K.,** Respiratory disease mortality among uranium miners, *Ann. N.Y. Acad. Sci.*, 271, 280, 1976.
3. **Hammond, E. C., Garfinkel, L., and Seidman, H.,** Some recent findings concerning cigarette smoking, in *Origins of Human Cancer*, Hiatt, H. H., Watson, J. D., and Winsten, J. A., Eds., Cold Spring Harbor Laboratory, Cold Spring Harbor, N.Y., 1977, 101.
4. **McCoy, G. D. and Wynder, E. L.,** Etiological and preventive implications in alcohol carcinogenesis, *Cancer Res.*, 39, 2844, 1979.
5. National Research Council, *Medical and Biologic Effects of Environmental Pollutants: Arsenic*, National Academy of Sciences, Washington, D.C., 1977.
6. **Hayes, R. B.,** Occupational exposure to chromium and cancer: a review, in *Review of Cancer Epidemiology*, Vol. 1, Elsevier/North-Holland, New York, 1979.
7. International Agency for Research on Cancer, *Monographs on the Evaluation of the Carcinogenic Risk of Chemicals to Humans*, Vol. 23, Lyon, 1980, 205.

IV. CARBONLESS COPYING PAPERS

Carbonless copying papers became infamous in the early 1970s when they were found to contain relatively high amounts of polychlorinated biphenyls (PCB) which to some extent could be transferred to the skin.[10]

The use of PCB as color-former solvent in these papers was quickly ended and prohibited in many countries. The ultimate disposal of the large quantities of PCB-containing papers in office archives has, however, not been regulated. It has been and is still possible to incinerate such papers in such a manner that PCB escapes combustion and is emitted into the environment with the stack gases.

Carbonless copying papers have recently been studied in connection with alleged discomforts from the exposure to such papers.[11,12] Except for an association with a particular (nondisclosed) color former for black color and accidentally high concentrations of biphenyl in some brands containing monoiosopropylbiphenyl,[12] none of the discomforts could be associated exclusively with carbonless copying papers. The symptoms: irritation in the nose and throat, itching of the face and eyes and on the exposed skin of the hands and arms may appear in any office environment with a high exposure to paper.

Carbonless copying papers contain a number of color formers as triphenylmethane, thiazine, oxazine, and xanthene dyes. These are potential mutagens but assays of extracts of carbonless copying papers have not shown any detectable response in the Salmonella mutagenicity test.[13]

V. LIGHTING

The incidence of malignant melanoma of the skin among caucasians has increased over the last several decades. Although increased exposure to solar ultraviolet (UV) light is considered to play a role, this has not been as firmly established as in the case for basal and squamous cell skin cancers.[14-17] The observation that office workers have a higher incidence than other indoor workers[17] led to a case-control study which indicated that exposure to fluorescent lighting at work increases the risk of malignant melanoma.[18]

It is known that fluorescent lamps radiate both in the UV-A and UV-B range,[19,20] but the average UV dose rate at ordinary illuminations is much less than that obtained in sunlight. Despite this, the hypothesis by Beral et al.[18] should not be overlooked for several reasons: the reported UV data for fluorescent lamps are probably not a representative selection of past and presently used lamps, the exposure is received during a long period compared with solar irradiation, the ratio of UV light to photoreactive (visible) light may be different for fluorescent tubes and sunlight.

The UV radiation from fluorescent lamps can be substantially decreased by the use of appropriate UV-absorbing glass in the lamps[19] and/or by using plastic diffusers in front of the lamps.[20]

VI. AIR POLLUTION

Indoor air pollution is now receiving increased attention as it is realized that a substantial part of any exposure to air pollutants is obtained in the indoor environment. Indoor air pollution has many origins such as combustion fumes from tobacco smoking, stoves, and heaters; radon daughters from radon escaping from the ground and building materials; volatile organics being released from building materials, indoor appliances, furnitures, and consumer products; and outdoor air entering the indoor environment.

Potential air pollutants which are specific for the office environment may originate from office materials and machines. Known offenders are solvents used for liquid photocopying toners, carbonless copying papers and felt pens. Some electric devices may generate ozone.

A. Mutagenic Colorants

The presence of mutagenic components in photocopying toners suggests that these compounds may become air pollutants in the copying process. Studies were therefore performed by collecting airborne particulate matter in the vicinity of photocopying machines with mutagenic and nonmutagenic toners as well as in control spaces.[13] These investigations did not show any detectable mutagenic response in the Salmonella assay that could be related to the copying machines. The finding is in accordance with the fact that dry toner beads have a size which will not yield airborne particulates. It is also unlikely that the mutagenic compounds which are polycyclic hydrocarbons and organic dyes will be released from the toner polymer in the heat fusing as these compounds have a low volatility.

The situation is physically different for office machines utilizing liquid ink. Liquid ink for a word processing machine has been found to contain mutagenic components which behave as aromatic amines.[7] High temperature laser fusing of colorants may also enhance the possibility of volatilization of organic compounds.

B. Tobacco Smoke — Passive Smoking

Passive smoking, i.e., exposure to the sidestream tobacco smoke and the exhaled part of the mainstream smoke, is a significant part of air pollution exposure.[21]

Sidestream tobacco smoke has a chemical composition that is different from that of mainstream tobacco smoke[22] which means that passive smoking cannot be expressed as cigarette equivalents except for reference to individual chemical compounds. Sidestream particulate matter, rather than the sidestream gas phase, is responsible for the irritating effects of tobacco smoke.[23] Sidestream tobacco particulates contain mutagenic compounds[24] as does the mainstream tobacco smoke.[25]

Office buildings are usually constructed with a ventilation system that conserves energy. Energy saving is often made by recirculation of the ventilation air. A study of such systems in two office buildings showed that the indoor pollution by particulate matter as measured by the Salmonella mutagenicity assay was higher than that of the outdoor ambient air and that the origin was tobacco smoke.[24] Energy conservation may also be made by the use of heat exchange systems. A study of such a system has revealed that the concentration of respirable particulates only decreased marginally in the nonsmoking spaces when smoking was allowed only in one part of the building, but that the concentration decreased by about 95% to outdoor ambient levels when smoking was completely prohibited in the entire building.[26] This result suggests that tobacco smoke particulates either spread faster from one room to another than the normally occurring ventilation rate or that the particulates are trapped in the revolving heat exchange tubes and are brought back into the building with the fresh air.

The only solution for nonsmokers to avoid passive smoking is to locate smokers in spaces that have ventilation systems which are completely separated from the nonsmokers' spaces and their ventilation systems.

VII. DISCUSSION

The office has been transformed from a workplace with plain papers, pens, and pencils into a complex environment with several new materials and machines.

The Throdahl[27] prediction that pencil graphite might be shown to contain hazardous impurities has partly been fulfilled in the sense that another carbonaceous structure, carbon black, used in office materials was shown to contain unwanted impurities in the form of mutagenic and, hence, potentially carcinogenic compounds. The mutagenic impurities in photocopying toners have largely been removed without apparently undue costs. It seems likely that similar impurities present in typewriter ribbons and carbon papers can be removed if the manufacturers address the problems.

The question can be posed whether or not the presence of mutagenic compounds in toners, typewriter ribbons, and carbon papers may be a hazard. If the evaluation is limited to the office workplace, the potential hazards are air pollution and transfer to the skin.

Air pollution by volatilization of the components is unlikely due to their high molecular weight and low volatility. A comparison can be made with a related compound, 2,3,7-trinitrofluoren-9-one (TNF) which has been used as photoconductor material in copying machines and printers. TNF is slowly abraded but no airborne concentrations have been detected near operating machines.[28]

The possibility exists that mutagenic compounds can be transferred to the skin in the handling of products and papers containing the colorants. The amounts which may be transferred are, nevertheless, low as most products, with the exception of carbon papers and their copies, are relatively non-smeary. The potential transfer to the skin from carbon papers and their copies is related to a similar problem for newspaper print which may contain mutagenic impurities.[3]

The evaluation of potential environmental hazards should, however, not be limited to a single compartment but involve all steps from the primary production to the ultimate disposal. It may be difficult to prove that all steps are sufficiently safe and particular obstacles involve the disposal of papers containing mutagenic colorants by paper recycling or by incineration.

Whereas there exist specific air pollution sources in offices such as organic solvents, the major air pollutants in the office environment originate from tobacco smoke. The office shares the tobacco smoke problem with many public places and private homes.

REFERENCES

1. **Löfroth, G., Hefner, E., Alfheim, I., and Møller, M.,** Mutagenic activity in photocopies, *Science*, 209, 1037, 1980.
2. **Rosenkranz, H. S., McCoy, E. C., Sanders, D. R., Butler, M., Kiriazides, D. K., and Mermelstein, R.,** Nitropyrenes: isolation, identification, and reduction of mutagenic impurities in carbon black and toners, *Science*, 209, 1039, 1980.
3. **Agurell, E. and Löfroth, G.,** Presence of various types of mutagenic impurities in carbon black detected by the Salmonella assay, in *Short-Term Bioassays in the Analysis of Complex Environmental Mixtures*, Vol. 3, Waters, M. D., Sandhu, S. S., Lewtas, J., Claxton, L., Chernoff, N., and Nesnow, S., Eds., Plenum Press, New York, 1983, 297.
4. **Zbinden, G.,** Interview with a toxicologist. II., *Trends Pharmacol. Sci.*, 3, 307, 1982.
5. **Sicé, J.,** Tumor-promoting activity of a n-alkanes and 1-alkanols, *Toxicol. Appl. Pharmacol.*, 9, 70, 1966.
6. **Horton, A. W., Eshleman, D. N., Schuff, A. R., and Perman, W. H.,** Correlation of cocarcinogenic activity among n-alkanes with their physical effects on phospholipid micelles, *J. Natl. Cancer Inst.*, 56, 387, 1976.
7. **Møller, M., Alfheim, I., Löfroth G., and Agurell, E.,** Mutagenicity of extracts from typewriter ribbons and related items, *Mutat. Res.*, 119, 239, 1983.
8. **Dabney, B. J., Mortelmans, K., and Wright, K. L.,** Mutagenicity testing of typewriter ribbons, *Environ. Mutat.*, 4, 411, 1982.
9. **Richold, M., Jones, E., Rodford, R., and Clarke, D. E. W.,** Mutagenicity of typewriter ribbons: use of the Ames test as a biological screen, in *Mutagens in Our Environment*, Sorsa, M. and Vainio, H., Eds., Alan R. Liss, New York, 1982, 479.
10. **Masuda, Y., Kagawa, R., and Kuratsune, M.,** Polychlorinated biphenyls in carbonless copying paper, *Nature (London)*, 237, 41, 1972.

11. **Göthe, C.-J., Jeansson, I., Lidblom, A., and Norbäck, D.,** Carbonless copy papers and health effects, *Opusc. Med. Suppl.*, 56, 1, 1981.
12. **Göthe, C.-J.,** Carbonless copying papers — a health problem? (in Swedish), *Lakartidningen*, 79, 336, 1982.
13. **Löfroth, G.,** Unpublished data.
14. **Magnus, K.,** Incidence of malignant melanoma of the skin in the five Nordic countries: significance of solar radiation, *Int. J. Cancer*, 20, 477, 1977.
15. **Teppo, L., Pakkanen, M., and Hakulinen, T.,** Sunlight as a risk factor of malignant melanoma of the skin, *Cancer*, 41, 2018, 1978.
16. **Houghton, A., Flannery, J., and Viola, M. V.,** Malignant melanoma in Connecticut and Denmark, *Int. J. Cancer*, 25, 95, 1980.
17. **Beral, V. and Robinson, N.,** The relationship of malignant melanoma, basal and squamous skin cancers to indoor and outdoor work, *Br. J. Cancer*, 44, 886, 1981.
18. **Beral, V., Shaw, H., Evans, S., and Milton, G.,** Malignant melanoma and exposure to fluorescent lighting at work, *Lancet*, 2, 290, 1982.
19. **Paulsson, L.-E.** UV-Radiation from fluorescent tubes Report 1979 -005, National Institute of Radiation Protection, Stockholm, 1979.
20. **Jewess, B. W.,** Ultraviolet content of lamps in common use, *Soc. Photo-optical Instr. Eng. J.*, 262, 55, 1981.
21. **Repace, J. L. and Lowrey, A. H.,** Indoor air pollution, tobacco smoke, and public health, *Science*, 208, 464, 1980.
22. Center for Disease Control, Involuntary smoking, in Smoking and Health, DHEW Publ. No. (PHS) 79-50066, U.S. Department of Health, Education, and Welfare, Washington, D.C., 1979, chap. 11.
23. **Weber, A., Fischer, T., and Grandjean, E.,** Passive smoking: irritating effects of the total smoke and the gas phase, *Int. Arch. Occup. Environ. Health*, 43, 183, 1979.
24. **Löfroth, G., Nilsson, L., and Alfheim, G.,** Passive smoking and urban air pollution: Salmonella/microsome mutagenicity assay of simultaneously collected indoor and outdoor particulate matter, in *Short-Term Bioassays in the Analysis of Complex Environmental Mixtures*, Vol. 3, Waters, M. D., Sandhu, S. S., Lewtas, J., Claxton, L., Chernoff, N., and Nesnow, S., Eds., Plenum Press, New York, 1983, 515.
25. **DeMarini, D. M.,** Genotoxicity of tobacco smoke and tobacco smoke condensate, *Mutat. Res.*, 114, 59, 1983.
26. **Nilsson, L. and Stenberg, R.,** Non-smokers' work environment (in Swedish), Report TULEA 1983:02, University of Luleå, Luleå, Sweden, 1983.
27. **Throdahl, M.,** The pencil problem -1990, *Chem. Eng. News*, 58(3), 5, 1980.
28. **Costello, R. J., Berg, S., and Klemme, J.,** Health Hazard Evaluation Report HETA 81-125-1029, HETA 81-352-1029, National Institute of Occupational Safety and Health, Washington, D.C., 1981.

Chapter 15

HAZARD AND SAFETY OF THE MODERN OFFICE ENVIRONMENT: ADMINISTRATIVE PERSPECTIVES

Ann E. Robinson

TABLE OF CONTENTS

I. INTRODUCTION

In recent years it has become evident that there is growing public awareness and concern about the effects of the working environment on workers' health and safety. In Ontario, the Royal Commission on Health and Safety of Workers in Mines Report of 1976 provided the major philosophical theme for the present occupational health and safety legislation. The report recommended the need for centralized and coordinated government auditing, it developed and advocated the concept of internal responsibility for health and safety, and stressed the need for openness in discussion and access to information. Since the passage of this legislation in October 1979, joint participation by both employers and workers in various aspects of occupational health and safety has been required and there is evidence of an increased awareness on the part of all parties in the workplace to practical aspects of health and safety in the workplace.

Recently the office, as a workplace has received considerable attention by government as well as the media with respect to potential hazards to workers' health. The increased sophistication of office equipment, emphasis on energy conservation with the implications for indoor air quality of reduced ventilation rates, workers' interest in their health and the right of all Ontario workers to information concerning health and safety have led to the devotion of greater attention and resources to these workplaces by occupational health and safety personnel in Ontario.

Compared with many industries, traditional office work has been considered safe, clean work but this is not to suggest that there are no potential hazards to health associated with office work. Recognition of the possibility of hazards to health and identification and definition of any such hazard represents both a change in our perception of, and a willingness to accept, the changing characteristics of the office environment. For example, constant sitting in poorly designed chairs, working in a "stuffy" atmosphere, or an office polluted by irritating fumes or in a stressful workplace with tight deadlines were all once accepted as "part of the job". Office workers no longer accept these conditions as being fixed — and rightly so.

Potential hazards to health of office workers are or may be associated with: (1) the design and layout of the office, the lighting and seating arrangements; (2) the presence of chemicals and certain physical agents including noise; (3) quality of the ambient air; and (4) job stress. The exact nature of the potential hazard or hazards will vary from office to office and may or may not include substances or agents known to be carcinogenic in man.

Concern about the indoor environment has increased, particularly since the emphasis on energy conservation has prompted some reductions in ventilation rates and often the intermittent use of ventilation systems. It has become apparent that potentially harmful chemical contaminants and bacteria may be recycled by faulty or poorly maintained ventilation systems. Chemical contaminants can originate from such common office products as solvents in correcting fluids, adhesives and cleaning fluids or fumes, from office machines or special printing processes. Adverse health effects may range from acute local to long-term effects after a latent period which may, perhaps, include cancer. For example, formaldehyde can be released from glues used on furniture made from particleboard, from building insulation, or fabric finishing agents. It may be of particular significance as an irritant in inducing sensitivity reactions or in affecting individuals with allergies or preexisting respiratory problems.

Increased levels of noise leading to worker complaints and discomfort may be a consequence of the open plan layout found in many offices. Fluorescent lighting in offices has been associated with malignant melanoma in an Australian study.[1]

From time to time some chemicals used in photocopiers have been identified as potentially harmful to health, for example the nitropyrene contamination of a carbon black used in a toner, and methanol used in spirit duplicating devices.

Carbonless copy paper became suspect because of reports that some workers suffer an allergic reaction to chemical sensitizers in the paper. The possible chemical hazards to health include formaldehyde, solvents, resins, and dyes. In addition, the substance of building materials, particularly if present in atmospheric dust, may be hazardous to health. Asbestos, which is regulated in Ontario, was extensively used in buildings to provide thermal and electrical insulation, and to act as a fire retardant.

II. OFFICE DESIGN

Seating and lighting are two important aspects of office design and those, if not adequate may cause discomfort and complaints and consequent stress.

Most office workers identify lighting as the most important aspect of their work environment. Yet how many workers can say that the lighting system in their office is "just right for them"? Studies have been conducted and tables published on how much light most individuals need under different working conditions. The effect of age on the preferred lighting levels is considerable, for the amount of light reaching the retina decreases with age. This factor together with the accumulation of dirt on fixtures or with the reduced output of light from fixtures over time may mean that the ideal situation is not maintained. Aging also makes the individual less resistant to glare. The reflective factors of walls, floors, and ceilings are generally considered at the design stage, as is the need for diffuse or indirect lighting. Light-colored matte finishes serve to reduce harsh contrasts between bright fixtures and dark background. Glare can result also from reflections off shiny surfaces. Involving the user in the selection and evaluation of new equipment, furniture, and office layouts may help to pinpoint areas of concern not readily evident to a designer less familiar with the specific work tasks to be performed.

There are many industrial jobs where workers are required to stand but where it is ergonomically practical to provide stools. In the office environment, most individuals do not have the option of standing or sitting, and the provision of a properly designed chair that allows for individual adjustment is more important. It is estimated that up to 40 min of productive time can be lost each day when an employee is not provided with properly designed seating. Office chairs should be adjustable in height and have an adjustable back rest. The chair should be adjusted to fit both the worker and the accompanying desk or machine. The material on the chair seat should be porous to allow normal body heat to dissipate.

In view of the time spent in the seated position, employees should be encouraged to stand or walk, particularly during rest breaks. The need for additional exercise will become increasingly evident in some clerical jobs as technology in the office increases, for the individual will file and retrieve electronically, rather than having to stand up to carry papers, etc., to another part of the office.

III. INDOOR AIR QUALITY

In the past, the effect on health of environmental pollution or of the effects of industrial pollutants in the workplace has received more attention. Lately, concern about the indoor environment, including that in offices and public buildings, often results from implementation of energy conservation measures.

By closing off exhaust ventilation and recirculating the indoor air, vapors from such common office products as solvents (in correcting fluids), adhesives and cleaning fluids, or fumes from office machines and special printing processes may increase exposure to air contaminants. The condition may, in some instances, be aggravated by inadequate maintenance of equipment.

Though concentrations may be low in comparison with guideline values, particular contaminants such as formaldehyde may affect hypersensitive individuals or persons with allergies, or preexisting respiratory problems.

It is not uncommon to find office equipment, such as photocopiers put into small rooms where air circulation is minimal or nonexistent. While the objectives of reducing the level of noise and removing the odors from the general office area may be achieved, air quality must be assured. In this instance, it is prudent to ensure that the area is properly ventilated, that the potential health effects of the chemicals used in the photocopying process are evaluated, and that noise is reduced to an acceptable level.

Criteria for the assessment of indoor air quality remain to be fully defined and many agencies have works in process. The origins of airborne contaminants merit further consideration since not only the equipment present but the fabric of the building, its heating and ventilation systems, and its furnishings and the lifestyle and habits of the workers are or may be significant. Much administrative work remains to be done even to establish minimum standards.

IV. OFFICE AUTOMATION

There have been many recent changes in technology. However, many cathode ray tube display units are being introduced into workplaces without either adequate consideration to ergonomic aspects of workplace design or the psychological effect upon those required to use the new equipment. There seems to be an inherent assumption that because the equipment is manufactured and sold for use in offices that it does not pose a hazard to health.

There are many examples where the "technology of the future" is being placed in yesterday's work environment. The result is documented examples of visual, general physical, and psychosocial discomfort. The most frequently reported symptoms relate to visual fatigue and include general discomfort, eye strain and blurring or irritated eyes. Symptoms of physical fatigue include almost daily pains in various body parts, including the neck, shoulder, legs, arms, and hands and wrists.

Most of these sources of discomfort can be reduced or eliminated by following the ergonomic principle of "fitting the task to the person".

The characteristics of the equipment, the work environment and the work task should be assessed to ensure that their design, placement, and usage are compatible with, and suitable for, the individual. Important factors are the readability of the display, the elimination of glare on the visual display terminal (VDT) screen and contrast glare from viewing a screen against a bright background, and the provision of furniture to ensure the correct posture while sitting at the keyboard and when typing from source documents. Variations in job demands and individual control of the tasks will help to maximize the comfort and sense of well-being of the operator.

Introduction of new technologies into the workplace often alters the organization and technical structure of work. There is a psychological effect on the individual who may no longer feel master of a job but rather mastered by it, especially if assessment of productivity is electronically monitored. Also, the recognition that simple, repetitive work, lack of interest, and management not fully utilizing the individuals' capabilities, can lead to feelings of powerlessness. Feelings of doing a job that is not important, of being isolated within the working environment, and of no longer being able to achieve personal goals and expectations, may lead to frustration and stress.

Psychological effects are known to arise from particular jobs and are often found in VDT users. Consideration of quality of the working life and practices such as job enrichment, job rotation, and the formation of work-sharing groups are effective ways of increasing the creative aspects of the job. In all situations, alleviation of the psychological or physical discomfort of the workers may be achieved by frank and open discussion, and involvement of the users from the planning and selection phases, through to implementation.

Another aspect of office automation merits some discussion. Apart from the use of some chemicals in some equipment, there has been much attention and contention during the past 3 to 5 years about radiation that is, or may be, emitted during operation of video display units. The Ontario Ministry of Labour experience based on measurements of many hundreds of such units is that no machine has been found to emit radiation in excess of guideline values. In most cases, emissions as assessed by "state of the art" instruments have been below the limits of detection. Measurements have been made at radiowave, microwave, and extra-low frequencies; of static and magnetic fields of X-rays; and on occasion, of ultraviolet emissions. The accumulated data from all tests carried out have consistently indicated that emissions are negligible and often not distinguishable from naturally occurring background. This experience is shared with that of experts in other jurisdictions and is consistent with the majority of current scientific opinion. Nevertheless, though poorly documented and mostly not substantiated by follow-up study, the reported association of a variety of adverse pregnancy outcomes with operation of a VDT is disturbing. Although there is, as yet, no sound evidence to support a cause-effect relationship between birth defects, miscarriages, and VDT operation, there is fear that there may be a risk to the embryo or fetus due to factors not yet identified or understood. It is not always appreciated that there is a natural incidence of adverse pregnancy outcomes. While there is a natural reluctance on the part of scientists to claim that any operation including use of a VDT is safe, this reluctance to make a categorical statement may create anxiety in the mind of the lay person. The importance of perceived concerns such as this is that, even if poorly founded in science, they may unduly aggravate stress reactions.

V. JOB STRESS

A stressful occupation can increase the risk of incurring a number of diseases and have physical consequences. An individual under stress may so damage or destroy relationships with friends, co-workers, or family members, that help from others is not forthcoming when needed most. When different jobs are compared, it is evident that rapid work pacing, especially by a machine, long working hours, and repetitive or monotonous work are the factors which lead to the highest levels of stress. However, harmful effects of job stress usually result from the combined effects of a set of job pressures and problems. Factors affecting job stress of clerical work include: (1) environmental factors (noise, lighting, temperature, and ventilation); (2) ergonomic problems (poor equipment, work station, and job design); (3) poor interpersonal relationships; (4) socioeconomic factors (low pay, discrimination, job insecurity, increased workload due to staff cutbacks); and (5) fear of new equipment and automation.

VI. ADMINISTRATIVE PERSPECTIVES

Administrators of occupational health and safety programs must be sensitive to both the perceived and actual hazards in the office. Ontario's Occupational Health and Safety Act is based on the premise of joint responsibility of employers and workers for health and safety in the workplace.

The Act contains a number of provisions which assist workers and employers in identifying and controlling workplace hazards.

The Act has many provisions relating to employer responsibilities to ensure the health and safety of workers including the provision, proper use and maintenance of equipment, materials and protective devices; providing information, instruction and supervision to protect a worker's health and safety; and taking every precaution reasonable in the circumstances for protection of workers. Workers also have responsibilities to report defective equipment or a contravention of the Act to the employer, to work in compliance with the Act, and to use protective devices provided by the employer.

A key issue is knowledge about what is or is not a hazard. As more attention is paid to occupational health and safety, office workers are examining their workplaces and questioning whether or not particular chemicals or equipment pose a hazard. It is a function of government to assist and audit the workplace to determine whether the provisions of the Act are being followed. It has been the Ontario Ministry of Labour's experience that there have been many enquiries concerning alleged occupational hazards in offices as distinct from industrial workplaces. Such enquiries have, on occasion been prompted by press reports. For instance, in 1980 there was an article describing the potential ozone hazard from xerographic process photocopiers. The Ministry of Labour received about 100 enquiries following publication of that article. An information sheet was prepared in consultation with the manufacturer of the equipment to provide guidance for minimizing exposure.

In addition, staff have checked ozone levels. No value was found that exceeded the recommended exposure criterion (TWA) of 0.1 ppm.

Under Ontario law, workers, including office workers, have the right to refuse to do work which they have reason to believe is unsafe. Complaints, usually of discomfort, from office workers usually result in checks on ventilation, assessment of pollutants, identification of chemicals, and the use of VDTs. Orders requiring remedial action are issued to employers were the Act or a regulation is found not to have been complied with. On occasion, changes may be recommended to increase worker comfort. The right to refuse unsafe work has been exercised rarely by office workers.

Joint health and safety committees which are required in many Ontario workplaces but not in offices provide a good mechanism for resolving health and safety concerns. Voluntary committees may be formed if both parties agree to do so and the Minister of Labour has the power to override the exemption and order that a committee be formed. With increasing office automation the dialogue between employer and worker associated with a committee may prove to be of increasing value to the wellbeing of office workers.

A major concern of all workers is the potential hazard to health of exposure to toxic substances. A requirement of the Occupational Health and Safety Act relates to the introduction of new chemicals into a workplace. Any employer who wishes to manufacture, distribute, or supply a new substance for use in a workplace, including an office, must first submit detail of its composition and properties. Data requirements for notification include the chemical structure, composition of the product, the chemical and physical properties, and sufficient toxicological testing results to allow assessment of the hazard to the health and safety of workers. Thirty-five new chemical agents were reported during the first three years of operation of the Ontario Act. The Act contains a two-tiered approach to controlling toxic substances. There are general provisions and the power to issue an order to prohibit or restrict the use of substances, or to order the adoption of administrative controls and work practices of engineering controls which are based on an opinion that the particular substance is likely to endanger the health of a worker.

The expert opinion may be based on information from any one of these sources: published data, a specially commissioned report, an assessment of the process, or an assessment of the properties of the substance used or intended to be used in the workplace. The expert opinion is to be provided at the expense of the employer and by a person with proper professional knowledge or qualifications. Those who manufacture, distribute, or supply chemical agents for use in any workplace in Ontario, including offices, are covered by this provision.

The Act provides for what are referred to in the Act as designated substances, "substances which warrant special attention through the setting of specific exposure limits with clear sanctions for non-compliance".

Some substances are subject to special regulations or designation. The selection of substances for designation is based on the historical association of serious illness or disease with exposure to the substance. In some instances this is supported by evidence of a serious potential risk of ill-health caused by persistent inadequacy of control measures in the workplace.

A comprehensive review of scientific literature, survey of programs adopted in other jurisdictions, and extensive discussion and public dialogue are included in the process used by the Ministry in the development of a regulation.

At present regulations in Ontario apply to lead, mercury, vinyl chloride, coke oven emissions, and asbestos. Three others, noise, isocyanates, and silica are nearly complete. In November 1982, the Ministry published a notice of intent to designate eight more substances; acrylonitrile, arsenic, benzene, cadmium, chromium, ethylene oxide, formaldehyde, and styrene.

The regulations require an assessment of the workplace and institution of any appropriate control program, including a medical program, where a hazard to health is found. The requirements are general in nature and detail is determined by the employer in consultation with the health and safety committee.

As we look to the future it is evident that the study and control of health hazards in the office environment is both complex and multifaceted. Administrators need to retain an open and flexible attitude and must consider what hazards are or may be associated with changes made in the fabric of the workplace, the ventilation system, and particularly with new equipment. A frank consultative process may best serve to protect the health of all workers including those in an office environment.

REFERENCE

1. **Beral, V., Shaw, H., Evans, S., and Milton, G.,** Malignant melanoma and exposure to fluorescent lighting at work, *Lancet*, 2, 290, 1982.

Chapter 16

THE PREVENTION OF OFFICE POLLUTION

R. Mermelstein, D. E. W. Clarke, C. J. DeMarco, J. C. MacKenzie, T. J. Roberts, and H. S. Rosenkranz

TABLE OF CONTENTS

I. INTRODUCTION

Sensitivity about the quality of the total environment is reflected in health and safety concerns regarding the office work environment. It is not surprising considering the radical change that is taking place in the office environment. Innovative technology, designed to improve productivity through constructive mechanization of office work is often viewed with suspicion. Such feelings are particularly acute when the technology is considered to have the potential of displacing additional people in a time of high worldwide unemployment. These trends, coupled with energy curtailment programs that lead to the recirculation of office ambient air and the architectural obsolescence of some office structures, have created an environment for conflict and concern. Xerox Corporation, aware of these issues, has consistently utilized safety as an essential constituent of the quality of its products. Office pollution can potentially be caused by any one or a combination of factors; people, places, and things or in more familiar terms, practices, facilities, and equipment/materials. We will demonstrate the care we take to prevent our equipment from becoming a source of office pollution.

II. POLICY

It is the policy of Xerox Corporation that all products and materials marketed worldwide by the Corporation meet the recognized standards for safety, health, and environment, and to assure that "good practice" is followed where no such standards exist or apply. In addition, Xerox products, materials, and practices must fully comply with the appropriate legislated standards. In instances where standards of different severity apply under various jurisdictions, our usual practice is to comply with the strictest standards multinationally.

III. PRODUCT SAFETY

All products manufactured by Xerox or for Xerox by others are "self-certified" that such a product is safe for use and service activities. As required by marketing, legal, and other considerations, Xerox products are typically submitted to and approved by agencies such as Underwriters Laboratories (UL), Canadian Standards Association (CSA), British Standards Institution (BSI), and Verband Deutscher Elecktrotechniker (VDE).

Each product program prepares a product safety plan which specifies the applicable design and service safety specifications. This plan lists the regulations to be considered and the standards to be met. Assessments are made for all possible hazards: electrical, mechanical, chemical, biological, and interactions between various possible hazards are considered. Extensive system testing is conducted under a variety of simulated field and stress conditions to verify that all the health and safety requirements have been met. Some of these tests are conducted by our internal test groups while others are performed by external test organizations.

Exposure to harmful light emission is prevented either by limitations of intensity levels or by the use of interlocks on platen covers. Laser products are designed and produced in such a manner that there is no customer exposure to laser radiation. Care is taken that service personnel are not exposed to harmful radiation levels while making adjustments according to established procedures. The product design, service procedures, materials, and special tools are all evaluated and safety approved prior to the start of production.

Space requirements for service operations are defined and temperature and humidity ranges are specified for certain products if extensive variations will affect performance.

A. Xerographic Process

In the xerographic imaging process, an electrostatic charge is applied to a photoconductive layer composed generally of selenium, or alloys of selenium with other elements. Following light exposure, the resultant latent image is developed with a finely divided electrostatic powder known as toner or dry imager. The toner is generally supplied to the latent image in the form of a developer mixture, composed of large carrier and small toner particles adhering to it, and is transported to it by a cascading or magnetically conveying process. The toner adhering to the imaged areas is transferred electrostatically to the paper and is permanently affixed to it by application of heat or heat and pressure. The residual toner on the photoconductor is removed, and the photoconductor is prepared for the next imaging cycle.

B. Toners and Carriers

Xerox toners are fine powders composed of plastics, colorants, and minor quantities of functional additives. Depending on the specific machine application either styrene-acrylic or polyester type polymers are the major component of the toner. Xerox does not use epoxy polymers in toners. In black toners, several different specialty grade carbon blacks are used as colorants, while for color copying various dyes or pigments are employed. The level of polycyclic aromatic hydrocarbon (PAH) trace impurities in the carbon black is strictly controlled and is exceedingly low. In some of the toners, after-treated carbon blacks are used hence the level of PAH impurities is even lower than above. During the toner manufacturing process, the carbon black/colorant and polymer are melt mixed and thus most of the colorant becomes encapsulated by the polymer.

Under normal operating conditions, the toners are entirely stable and no significant amount of decomposition takes place. They merely flow and adhere to the paper upon the application of heat or heat and pressure depending upon the specific machine.

Xerox carriers are based on special grades of sand, glass, or steel or ferrite type materials, ranging in size from about 100 to 600 μm in size. They are generally coated with a small amount of special polymer to achieve the desired functional behavior in the copier or duplicator.

C. Photoreceptor

The photoreceptors or photoconductors used by Xerox are principally composed of amorphous selenium. A number of photoreceptors use a selenium alloy which contains a very low level of arsenic (less than 0.5%), while others use an alloy based upon arsenic triselenide. In selected applications, the photoreceptor may contain tellurium. The photoconducting layers are very tightly bonded to a substrate which is an alloy of aluminum for rigid substrates or drums, while electroformed nickel is used as the substrate for flexible application or belts. One of our more recent products, the 1075 system, uses a new photoreceptor consisting of a three-layer photoconductor on a polyester film base. It contains some selenium and a new organic photo conductor.

Because of the nature and composition of the photoreceptor, it is a requirement of Xerox that all used photoreceptors be returned to the company for proper disposal. Both the substrate and the photoconductor are recycled to conserve materials. Scrap materials containing arsenic or selenium are disposed in licensed hazardous materials disposal sites.

D. Emission Measurements

Machine emission measurements are performed in a closed test chamber of approximately 28 m^3 volume and having polytetrafluoro ethylene (PTFE) covered walls. The environmental conditions are generally 22.5 $\pm$ 2.5°C, 45 $\pm$ 5% relative humidity (RH), and ventilation is adjusted to 0.5 air change per hour.

Ozone measurements are made by direct monitoring techniques, either chemiluminescence or UV spectrophotometry procedures are used. Instruments are calibrated against chemical or ultraviolet standard. The detection limit is 0.001 ppm.

For the measurement of airborne metals or metal salts, a combination of sampling trains consisting of filters and solutions are used. For arsenic, selenium, and tellurium, particulates are collected on a sampling train consisting of a filter, two liquid bubblers, and a gas trap. The method is designed for sampling machine emissions over a period of several hours. The detection limits for arsenic, selenium, and tellurium using this methodology are 0.006, 0.03 and 0.001 $\mu g/m^3$ in a 1.0 m^3 sample, respectively. This sampling methodology may also be applied to sampling for the above elements in other situations, provided sampling rate and volume are adjusted to suit the expected airborne concentrations.

For normal operator position sampling, the filter holders are set at a height of 1.2 m above the floor and 0.3 m away from the machine. The other traps are set as close together as possible after taking steps to ensure that they do not hinder the necessary machine running operations. Similar precautions are taken when the sampling is at the service engineer or machine exhaust(s) positions. In the latter case, the filter holder is set flush against the machine exhaust positions.

While sampling runs are in progress, both static and dynamic blanks are taken. Static blanks are filters and Drechsel bottles which are retained on the laboratory bench throughout the sampling run. Their purpose is to provide information on background levels of the elements measured (arsenic, selenium, tellurium) in the filter papers or reagents and any contamination due to handling and analysis. Dynamic blanks are trains set up like sampling trains and run in parallel with the sampling runs in a location similar to, but remote from, the sampling position. The purpose is to provide information on the airborne background levels of the elements of interest.

For the determination of arsenic, selenium, and tellurium, the combined hydride generation atomic absorption spectrophotometry techniques are used. In the arsenic determination, a modified hydride generation technique having higher sensitivity than the conventional method is employed.

E. Emissions — Sources and Causes

Ozone is primarily produced in xerographic copiers and duplicators by the corona discharge of the various corotrons. UV emissions from document exposure lamps are extremely low, hence ozone generation by this means is insignificant. Since corotrons are not energized in the standby mode, significant ozone generation can only take place when the equipment is actually making copies. Some Xerox equipment requires ozone limiting devices. The Xerox Ozone Management Program requires that machines situated in locations that do not meet either space or environmental (temperature and humidity) requirements must be equipped with a filter to reduce ozone to an acceptable level. Some machines are equipped with ozone filters at the factory while alternatively others may be retrofitted with such a device at the placement site; the combined heat and ozone emissions may also be controlled via ducting. Ozone is unstable and changes to oxygen very rapidly. All Xerox products, operating normally over an 8 hr working day under the minimum specified environmental conditions meet the worldwide standard of 0.1 ppm time weighted average (TWA) concentration.

F. Emissions — Results and Discussion

During the normal copying/printing operation, extremely small quantities of the surface of the photoconductor may be abraded. In addition, the various operations involving toner result in some of the toner becoming airborne. Most of the materials are retained within the xerographic equipment in filters, traps, etc., but small portions are carried out of the machine. As part of the normal development activities, Xerox strives to keep these emissions to a minimum, consistent with safe and reliable operation of the equipment.

Table 1
TYPICAL EMISSIONS FROM XEROX PRODUCT FAMILY

Substance concentration	Permissible limit[a]	Low volume: Xerox 2830	Low volume: Xerox 3100	Mid volume: Xerox 1075	Mid volume: Xerox 5600	High volume: Xerox 9200	High volume: Xerox 9500
Ozone ppm[b]	0.10	0.05	0.045	0.010	0.045	0.055[c]	0.066[c]
Arsenic mg/m³ [d]	0.010	0.00001	0.00002	N.A.[f]	0.00001	0.00004	0.00020
	0.200[e]						
Selenium mg/m³ [d]	0.200	0.00001	0.00007	N.A.	0.00001	0.00020	0.00003
Tellurium mg/m³ [d]	0.100	0.00001	N.A.	N.A.	N.A.	N.A.	N.A.
Toner mg/m³ [d]	15.00	0.020	0.060	0.04	0.100	0.020	0.100
	10.00[g]						
	5.00						

[a] U.S. Government Standard-Permissible Exposure Limit. The maximum permissible exposure an employee may experience from airborne substances averaged over an 8-hr period of time as mandated by OSHA.
[b] ppm = Parts per million; at minimum Xerox siting requirements.
[c] Calculated value.
[d] mg/m³ = Milligrams per cubic meter.
[e] American Conference of Governmental Industrial Hygienists (ACGIH) limit.
[f] Not applicable.
[g] The value of 15 mg/m³ is the OSHA limit while the value of 10 mg/m³ is the American Conference of Governmental Industrial Hygienists limit. Both OSHA and ACGIH use a 5 mg/m³ as the respirable limit.

The results of typical measurements of ozone, arsenic, selenium, tellurium, and toner emitted from six widely distributed machines are shown in Table 1. The values listed are based upon measurements made in the instrumented test chamber, and provide a fair indication of what an individual would be exposed to operating the equipment 8 hr a day. Based upon our information, such a high degree of equipment utilization is unusual and is encountered rarely and then only in high volume copiers.

A comparison of the measured values with the applicable U.S. standards shows all of them to be well within the required limits. In many cases, the actual measurements are at or near the limit of detection. Thus, the measured values for arsenic are 250 to 1000 times less than the standard and same values for selenium are 1,000 to 10,000 less than the standard. The values listed for toner/dust are a composite of airborne dust, paper dust/fibers, and toner. Even if all of the toner/dust were of respirable size range, which is not the case, the measured values are 50 to 250 times less than the permissible level. Therefore, these emission levels do not represent any risk to the operators of the equipment and the public at large.

IV. MATERIALS SAFETY

To assure full compliance and implementation of the safety policy, health and safety considerations are an essential element of the product and material design processes. Objectives are defined, criteria agreed upon, and progress is measured during the evaluation of a product or material.

With regard to the various xerographic materials, we have historically performed the appropriate acute toxicity tests (ingestion, inhalation, sensitization, etc.). When appropriate, the results of longer time period exposures are evaluated. In view of the recent advances in the field of genetic toxicology, for a number of years we have used a battery of tests as potential predictors of longer range effects.

In an unrelated activity, Scandinavian scientists using the Salmonella assay detected a minor amount of mutagenic activity in the extracts of some xerographic copies and toners.[1] The cause of the observed mutagenicity was due to nitropyrenes, which were found to be a trace impurity the particular carbon black used to fabricate the specific toners.[2] As a consequence of cooperative efforts in materials analysis and process control by Xerox and our vendor, these impurities have been reduced to such a low level that they are not detectable in the affected toners. The corrections to this problem were implemented in the spring of 1980. Of course, none of the Xerox toners or other materials show mutagenic activity now in the Ames Salmonella assay.

As a consequence of the above findings, we have reexamined the safety information on a number of Xerox materials and implemented corrective actions when appropriate.[3] We have also instituted strict mutagenicity specifications on several raw materials. Specifically with regards to carbon blacks our practice is to utilize Best Available Technology (BAT) to minimize mutagenic impurities. A specification limit of 1 ppm max total was established to control the permissible quantity of extractable "specified PAH" or other mutagenic impurities. In this context, "specified PAH" or derivative refers to chemicals which are known or alleged to have carcinogenic response in animals.

V. MATERIALS SAFETY ASSESSMENT

In a broader context, we have developed and implemented a Materials Safety Assessment Standard in order to assure that a comprehensive and cost effective safety evaluation of all materials and products takes place in accordance with Xerox policy. The standard is applicable to the Corporation and its subsidiaries worldwide. The materials safety assessment process is based upon published accounts which are used by several other organizations.[4-7]

For the purposes of the standard, the term material is used broadly and includes raw materials, intermediates (independent of state of isolation), fugitive or processing materials, finished material(s) or products, and waste products or byproducts.

A. Preface

The materials safety assessment process requires the identification, generation, collection, integration, and evaluation of information. The strategy adopted to achieve even the minimum requirements of this standard can be both complex and expensive. The intention is to outline the minimum requirements and allow for both judgment and expertise to assure the flexibility needed in acquiring the necessary information. Whenever possible, definition of the type of information required rather than the specific test assays has been listed. Since some areas of the field of toxicology are developing very rapidly, it is expected that currently preferred test systems or assays will undergo change with time.

B. Operating Principles

The following principles will be utilized in the safety assessment of materials:

1. All products and materials marketed worldwide by Xerox Corporation will meet the recognized standards for health, safety, and the environment, and assure that "good practice" is followed where no such standards exist or apply.
2. Materials will not be used or marketed which expose Xerox employees, customers, the general public, or the environment to potentially hazardous conditions.
3. Materials having the lowest possible toxic potential are preferred. Materials exhibiting or suspected of producing a potential toxic effect may be used, provided adequate safeguards exist and less toxic alternatives are not available.
4. Research and development use of materials must comply with "Prudent Practices for Handling Hazardous Chemicals in Laboratories" (see Section VI).

5. Facilities design, including engineering controls are the primary method of controlling employee exposure to chemical and physical agents. Personal protective equipment, work practices, and/or administrative controls are utilized only when engineering controls are not feasible or as a supplement to engineering controls.
6. Materials reuse and recovery are incorporated into the design process to conserve valuable materials and resources. Engineering controls are then utilized for the remaining air emissions, solid waste, and wastewater effluents before discharge to the environment.
7. All external contract sources/organizations involved in the generation of information required for the safety assessment of materials must be qualified/validated so that they meet the highest standards for quality and reliable performance consistent with the requirements of OECD or similar guidelines.
8. In the areas where expert judgment is required or when substantial differences of opinion exist among internal knowledgeable personnel with regard to the assessment process, the advice and counsel of competent and recognized consultants is expected.
9. The scientific publication of environmental health and safety information is encouraged while protecting proprietary information.
10. Reprographic Business Group's Operational Environmental Health and Safety is recognized as the lead multinational organization for the assessment of materials and product safety.
11. Operational Environmental Health and Safety shall procure and maintain the necessary competencies to assure the orderly implementation of the Materials Safety Assessment Standard.
12. Operating groups with limited "safety know-how" may obtain the required safety information by purchase of services (or other arrangements) from the Operational Environmental Health and Safety Organization or through contract with qualified external safety test laboratories.
13. The function of the external safety test laboratory is to perform specific laboratory assays and/or measurements. It is not their function to perform safety assessment.
14. In order to minimize the risk of erroneous test results and due to the special proprietary nature of some of the materials and/or processes, it may be necessary, in certain instances that several different contract sources furnish the required information.

C. Assessment Process and Recommendations

Both a Preliminary and a Formal Safety Assessment shall be required. The description of the assessment process is as follows: The objective of the Preliminary Safety Assessment is to review the basic safety information data base on the material and assess the degree of risk that may be involved. This process requires an identification of the potential hazards, an estimation of risk involved, and the potential means of containing or limiting such risk. This process must also arrive at a determination regarding the adequacy of the available safety information data base to permit a product commitment. If significant potential health risks have been identified, it may be necessary to delay product commitment pending resolution of major issues or concerns.

It is anticipated that in most instances, some additional information will be required which can be generated in the interval between the Preliminary and the Formal Safety Assessments. In exceptional situations, it may be necessary to obtain additional safety information based on the outcome of the Formal Safety Assessment. This assessment standard and schedule is designed to avoid adverse information at a later date and therefore, assure timely market introduction.

The objective of the Formal Safety Assessment is to review the entire safety information data base on the material and certify that it is safe and ready for market introduction. Particular attention should be focused on the information derived from supplemental tests which were performed between the Preliminary and the Formal Safety Assessment. Any new information, whether originating from supplemental tests defined at the Preliminary Safety Assessment or arising from external sources, should be critically assessed and so verified that the total safety information data base supports a market introduction action.

It should be understood that the information requirements of both the Baseline and Supplemental categories (see Section V. H. 2) represent a trade-off between the time and cost of a wide variety of health and environmental tests and a degree of manageable risk. The nature and extent of the testing should be commensurate with the degree of potential risk entailed. As new information becomes available a reevaluation of the entire process should be carried out in order to improve and maintain an up-to-date overall risk assessment.

The critical elements of the assessment process are

1. *Hazard identification*: the objective of this first stage in risk assessment is to establish qualitatively whether a potential health hazard exists. Information is developed and evaluated according to the evidence concerning its potential health consequences.
2. *Risk estimation*: once a potential health hazard has been identified and evaluated, it is necessary to quantify the potency of the material and the factors relevant to human and environmental exposure. By acquiring industrial hygiene and environmental information and relating it to physio- and biochemical effects the potential risk is estimated qualitatively in the best possible manner. Only rarely is there sufficient evidence available to enable a risk estimation to be expressed numerically, (e.g. 1 in 10 risk of cancer on exposure of 10 ppm). The question thus will become one of expert judgment rather than that of mathematics. Hence the advice and expert opinion of consultants in the various fields of toxicology is strongly encouraged.
3. *Risk containment*: information from the risk estimation stage is considered in the context of the following necessary factors: (a) the means available to limit exposure; (b) the technological feasibility and cost of implementing administrative, engineering, and/or protective equipment control measures; (c) the existence of alternative less hazardous materials or processes; (d) the availability of analytical and monitoring methods to measure exposure; and (e) the social, economic, and legal advantages and penalties. The end result of this stage is the development of recommendations governing exposure conditions, the special handling of the material, and any other measures necessary to ensure that the perceived risk is controlled to a suitably low level by means which are technologically feasible, practical, cost effective, and which take into account the social and economic consequences of the proposals.

The attention of all participants in the assessment process is directed to two additional aspects which nevertheless provide critical dimensions to the ethical and legal framework. The risk assessment process requires a substantial degree of flexibility and judgment. Thus, particular attention must be placed upon the consistency, objectivity, and ethics of the assessment process. The Xerox safety professionals who control the information have a special obligation to the safety of other employees, the consumers, and the public. Attention is also directed to the Xerox Corporate Policy R&D 011 regarding notification of substantial risk. This policy establishes provisions for corporate reporting of substantial risks of injury to health or the environment to the Environmental Protection Agency (EPA), as required by the Toxic Substances Control Act (TSCA) in the U.S. and as may be required by similar laws in other countries where Xerox operates.

D. Timing of the Assessment Process

Early acquisition of selected EH/S information during the research and development stage

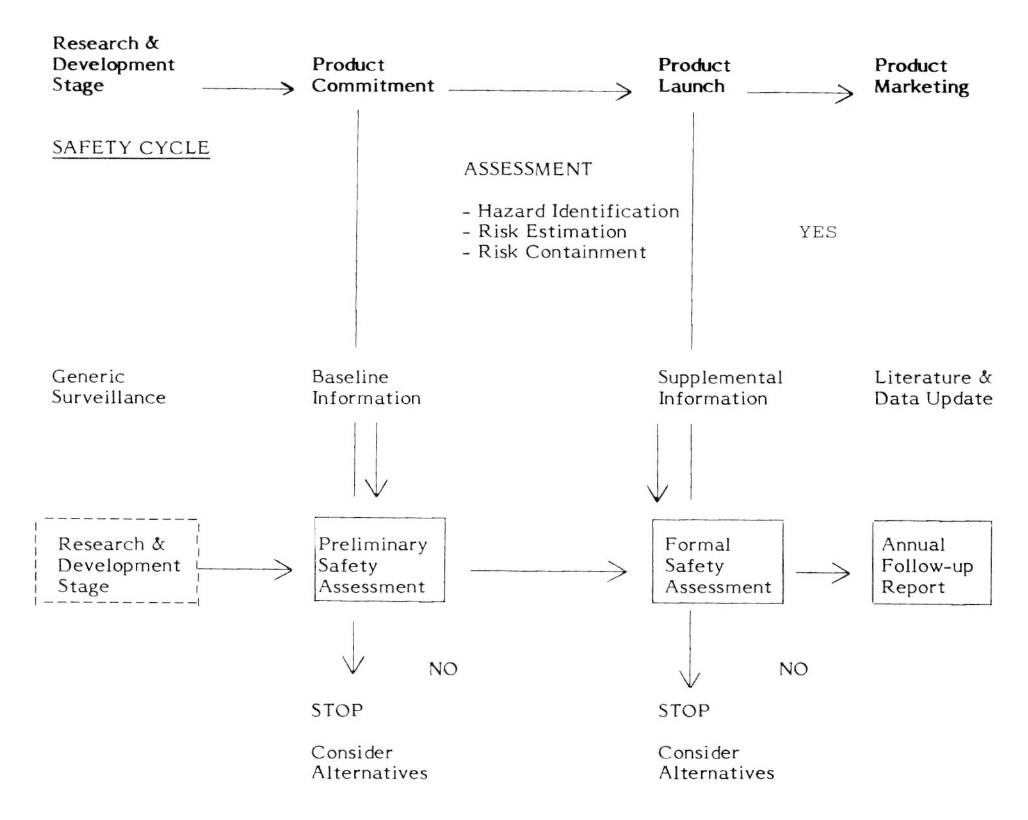

FIGURE 1. Process illustration.

of technology activities is encouraged. The relationship between the Product/Program Development cycle and the Materials Safety Assessment process is illustrated in Figure 1.

1. Preliminary Safety Assessment

A preliminary safety assessment must take place prior to a commitment to develop a product.

2. Formal Safety Assessment

A formal safety assessment must take place prior to introduction of the material/product to the marketplace. The precise timing of the assessment will depend on a number of factors, including the degree of complexity of the material/product, the nature of the issues analyzed, etc. The assessment shall take place early enough to permit corrective actions to be taken, should that be required.

E. Responsibilities

1. Corporate Environmental Health & Safety: (a) provide independent safety assurance of materials to corporate management; (b) integrate significant external factors and take appropriate action to protect broad company interests; and (c) maintain surveillance of EH/S aspects of the research and development stage of technology activities.
2. Operational Environmental & Safety: (a) assure the safety of all materials in accordance with corporate requirements; (b) maintain surveillance of and acquire selected EH/S

information during the research and development stage of technology activities: (c) provide safety approval for product commitment and market introduction; (d) assemble and integrate safety information, and keep corporate EH/S apprised; (e) assume lead role in data evaluation and assessment and conduct the Preliminary and Formal Safety Assessments; (f) provide in a timely manner a report on the status and recommendations arising from both the Preliminary and Formal Safety Assessments; (g) provide Material Safety Data Sheets (MSDS), Product Safety Data Sheets (PSDS), and health and safety aspects of service training instructions at the time of the Formal Safety Assessment; and (h) maintain on-going literature and test information surveillance (as new information becomes available, a reevaluation of its impact on any prior assessment is required).

3. Operating Strategic Business Unit/Subsidiary Management: (a) ensure Materials Safety Assessment is carried out by Operational Environmental Health and Safety; (b) participate in the Preliminary and Formal Safety Assessment; (c) respond to requirements arising from the assessment process; (d) acquire and disseminate health and safety related service training instructions; (e) alert Operational EH/S on critical field safety issues; and (f) advise Operational EH/S of new materials development anticipated material changes, including product and service applications.

F. Categories of Information

The following categories of information shall be considered: (1) effects of exposure on human health; (2) effects of exposure on the environment; (3) magnitude of health exposure; (4) magnitude of environmental exposure; and (5) external factors.

G. Evaluation Criteria

The following criteria shall be utilized in determining the magnitude and effects of exposure on both humans and environment. In each group, three arbitrary levels are considered:

1. Criteria for Health Effects
   a. Gross toxicity (LC_{50} and/or LD_{50} mg/kg)
      - >500
      - 50—500
      - <50

   b. Immediate effects
      - None
      - Reversible
      - Irreversible

   c. Prolonged effects
      - None
      - Reversible
      - Irreversible
2. Criteria for Environment Effects
   a. Compatibility with waste treatment
   b. Aquatic toxicity
   c. Bioconcentration potential
3. Criteria for the Magnitude of Health Exposure (Annual and cumulative to the lifetime of the program)
   a. Production/use quantity
      - 10^3 kg
      - 1—100 × 10^3 kg
      - 100 × 10^3 kg

- b. Number of people exposed
  - 1000
  - 1000—1,000,000
  - 10^6
- c. Duration of exposure
  - 200 hr/year
  - 200—2000 hr/year
  - 2000 hr/year
- d. Type of populations exposed and description
  - single (e.g., manufacturing, R & D)
  - two (e.g., employee + consumer)
  - three (e.g., employee + consumer — e.g., predominantly female — + general public)

4. Criteria for the Magnitude of Environmental Exposure
   - a. Discharge quantity
   - b. Number of discharge sites
   - c. Frequency of annual discharge
   - d. Dwell time after discharge
5. External Factors
   - a. Regulatory
   - b. Legal
   - c. Societal
   - d. Public Relations
   - e. Industrial Relations
   - f. Personnel Relations
   - g. Other

H. Baseline and Supplemental Information

The baseline information shall constitute the minimum requirements of the Preliminary Safety Assessment. The baseline information must be acquired and assembled on all new materials and products or on existing materials when there is a significant new use or change in sourcing or manufacturing process. The previously assembled information shall then be compiled and evaluated in accordance with the following outline.

In instances of certain materials related changes, (e.g., change in vendor, materials replacement) frequently it will be sufficient to reference the prior "Baseline Information" data base and indicate which portions of the information are essentially unchanged. In these cases, the focus shall be on aspects which differ from prior experience and an update or revision of time dependent information shall be provided.

1. Material, product application and function
   - a. Nature of the material
   - b. Method of use
   - c. Relevance to existing and/or currently used and approved materials
2. Estimates of quantities to be used and disposed
   - a. Estimated annual and total amount over life of the program
   - b. Quantities of the material to be utilized, purchased, or manufactured and disposed by Xerox or others
   - c. Anticipated methods of disposal
3. Estimates of exposures
   - a. Categories of population to be exposed, (in-house, R & D, manufacturing, service personnel, customer, the general public, etc.)

   b. Anticipated level and duration of exposures as a function of time
   c. Estimated number and types of people in each of the above categories who may be exposed
4. First order estimates of activity
   a. Structure of the material and related analogs/homologs
   b. Immediate and prolonged potential health effects
   c. Reversibility or nonreversibility of such effects
   d. Toxicity rating of the material (extremely toxic, moderately toxic, practically nontoxic)
5. Literature survey
   a. Comprehensive literature survey regarding the current use, properties, and toxicology of the material or its components
   b. Methods and techniques for analytical measurement including analysis in biological systems
   c. Information regarding facilities design, engineering controls, safe workplace procedures, and protective equipment
6. Physicochemical data
   a. Physical state and fundamental properties of the molecule or chemical
   b. Molecular weight
   c. Melting or boiling point
   d. Solubility
   e. Corrosivity
   f. Reactivity
   g. Ignitability/explosibility
7. Health and environmental data
   a. Acute toxicity
      i. Oral, inhalation LD_{50}
      ii. Skin and eye irritation
   b. Mutagenicity — the Ames *Salmonella* assay is viewed as a primary screening tool for chemicals; instances of a positive response or an adverse outcome do not automatically disqualify a material, but they necessitate a series of follow-up actions which are frequently time consuming and costly; thus in many cases, the use of an alternate material may be simpler and preferable
      i. *Salmonella typhimurium* — (at least 5 tester strains and in the absence and presence of metabolic activation)
      ii. *Escherichia coli*
      iii. At least one mammalian cell assay — the choice of the test system selected should be governed by the nature of the test validation data, literature study, i.e., have homologs been tested in the particular assay, and the chemical structure of the material, and whether one of the systems are specified or required by regulatory agencies
   c. Biodegradability BOD/COD ratio

Supplemental information from any of the following categories may be required in order to complete either the Preliminary or Formal Safety Assessment.

1. Toxicity tests
   a. Repeated skin application/sensitivity
   b. Genetic toxicology
   c. Subchronic feeding or inhalation
   d. Reproductive toxicology

  e. Metabolism studies
  f. Transformation assays
  g. Carcinogenicity assays
2. Environmental tests
  a. Octanol/water partition coefficient
  b. Acute fish toxicity tests
  c. Extraction Procedure Toxicity
  d. Air Quality Modeling

I. Documentation

The following written documentation is required: (1) a report on the status and recommendations from the preliminary and formal safety assessments shall be filed with Corporate EH/S within 15 days of the completion of such assessments; (2) both MSDS and PSDS as well as the health and safety aspects of service training instructions must be completed within 30 days of the Formal Safety Assessment but prior to market introduction; and (3) an annual summary to the record shall be provided, indicating that a review and evaluation of all new information has been carried out, and evaluated in the context of the prior assessment.

J. Information Dissemination

Summaries of the safety information are prepared and published in form of Data Sheets. The Product Safety Data Sheet (PSDS) contains information about the electrical, mechanical, and environmental requirements of the product as well as the emissions of the system into the environment in which it is located. Similarly, Materials Safety Data Sheets (MSDS) identify the material, provide information regarding physical properties, toxicology, emergency first aid, and any other specialized data. These data sheets are available to all who request it, be they employee, customer, government, or other regulatory organization. These documents are periodically reviewed, updated, and thus reflect new information as it becomes available.

VI. EMPLOYEE SAFETY

A. Factory — Xerox Site

At the various Research, Development, and Manufacturing sites, comprehensive and up-to-date safety programs are in place. All new facilities and processes must be approved by the appropriate safety engineers prior to start-up. Operations involving potentially hazardous materials are thoroughly evaluated to assure adequate controls and procedures are in place to protect employees and facilities. Periodic industrial hygiene audits are carried out to verify that processes are operating within defined limits and employees are not exposed to harmful substances. Routine biochemical assays in blood and urine of personnel in our Research, Development, and Manufacturing Operations, where potential for exposure could be much greater than actual field conditions, have all indicated levels within normal limits.

An occupational health information data system is being implemented to detect health trends that might be related to the workplace environment. This system will provide Xerox Corporation with the latest methodology in assessing and maintaining the health of its workforce.

B. Technical Representatives — Customer Site

1. Responsibilities

Xerox Corporation employs more than 10,000 Technical Representatives in North America. These employees are responsible for the service and maintenance of Xerox products at customers' sites. Their responsibilities include the installation and relocation of machines, routine maintenance, and repairs which are primarily electrical or mechanical in nature. During these operations, the Technical Representatives are exposed to both materials contained in the machines as well as specialized materials used in servicing the equipment.

Due to the nature of these tasks, it is assumed that the exposure of the Technical Representatives to these chemicals is several times higher than that of any customer. The specific activities which may involve exposure to the xerographic materials include photoreceptor cleaning or polishing, change of the developer, cleaning of the drum module and corotrons, replacement of filter bags and cleaning of the toner catch tray, and cleaning of the fuser area including addition of fuser oil and replacement of the fuser wicks. Efforts have been made and are continuing to keep exposures to a minimum, e.g., in polishing of the photoreceptor it is mandatory that a wet cleaning process be used to keep dust to a minimum.

2. Industrial Hygiene Measurements

The measurements and results discussed below were carried out by an independent external firm. The investigation was performed in three segments, the first and third of which required calculations of TWA exposures. The first part of the study consisted of monitoring of Technical Representatives exposures under controlled environmental conditions. The site was a Xerox internal training/service facility and the various tasks were evaluated under "worst-case" ventilation conditions by deliberate closing of doors and windows and turning off the HVAC systems. These conditions were selected in order to mimic what might be worst case condition in the field at a customer site. During the second phase of the testing 15 to 20 Technical Representatives of the "internal branch" were fitted with sampling equipment which they wore throughout the day for at least 2 days. Each of these employees was provided with an activity log to describe clearly the tasks accomplished. These activity logs combined with the machine records provide a good indication of the conditions of the equipment, nature of site ventilation, and personal exposures. The third and final part of the study involved the servicing of what is considered to be the dirtiest machines in order to approximte potential worst case exposures. In this case, the Technical Representatives were requested to clean the dirtiest machines available at the Webster Refurbishing Center. Several of the Technical Representatives indicated that they would not have serviced a machine in such a deplorable condition at a customer site, but would have arranged replacement equipment for the customer. Therefore, we feel reasonably confident that the servicing of these equipment represents worst case exposures. With respect to the "dust-toner" measurements, the values shown reflect the total pyrolizable particulate matter, which is only in part toner.

C. Results and Discussion

The results for all of the samples collected and measurements made without exception indicated TWA contaminant concentrations below the most stringent applicable limits. The standards referenced were Occupational Safety and Health Administration (OSHA), the American Conference of Governmental Industrial Hygienists (ACGIH), and Xerox internal exposure limits.

In several instances, the actual measurements are below or near the limit of detection. The median TWA measurements for arsenic, selenium, total-dust, and isopropyl alcohol for each of the three segments of the study are shown in Table 2. The range of TWA exposures for each of the above chemicals is shown in the last column of Table 2. A comparison of the measured values with the strictest applicable standards shows all of them to be well within the required limits. Thus, the highest arsenic and selenium concentrations measured were three orders of magnitude below the current OSHA permissible exposure limit (PEL) and ACGIH threshold limit value (TLV), respectively. The highest concentration of iopropyl

Table 2
TYPICAL EXPOSURES MEASURED FOR XEROX TECHNICAL REPRESENTATIVES

Substance concentration	Permissible limit[a]	Mean TWA exposures calculated I	II	III	Range of TWA exposures calculated: composite
Ozone ppm[b]	0.10	N.D.[c]	N.D.	N.D.	N.D.
Arsenic mg/m^3 [d]	0.010	0.00005	0.00005	0.00003	0.00002-0.00020
	0.200[e]				
Selenium mg/m^3 [d]	0.200	0.00026	0.00014	0.00009	0.00004-0.00040
2-Propanol ppm[b]	400	1.6	3.8		0.11-5.4
Toner mg/m^3 [d]	15.00	0.33	0.30	0.38	0.09-0.94
	10.00[f]				
	5.00				

[a] U.S. Government Standard-Permissible Exposure Limit. The maximum permissible exposure an employee may experience from airborne substances averaged over an 8-hr period of time as mandated by OSHA.
[b] ppm = Parts per million; at minimum Xerox siting requirements.
[c] None detected.
[d] mg/m^3 = Milligrams per cubic meter.
[e] American Conference of Governmental Industrial Hygienists (ACGIH) limit.
[f] The value of 15 mg/m^3 is the OSHA limit while the value of 10 mg/m^3 is the American Conference of Governmental Industrial Hygienists limit. Both OSHA and ACGIH use a 5 mg/m^3 as the respirable limit.

alcohol measured was approximately two orders of magnitude below the OSHA PEL of 400 ppm. Similarly the highest dust-toner concentration measured was a full order of magnitude below the Xerox internal limit of 5 mg/m^3. Therefore these measurements are indicative of minimal exposures and do not represent any risk to the Xerox Technical Representatives or the customer, and thus to the public at large.

VII. SUMMARY

It is a fundamental principle of Xerox Corporation to assure that our products are safe and do not in any way represent a concern to our customers or our employees. While the above results document only a small part of our safety activities, they demonstrate how seriously we view and exercise our safety responsibility.

REFERENCES

1. **Löfroth, G., Hefner, E., Alfheim, I., and Moller, M.,** Mutagenic activity in photocopies, *Science*, 209, 1037, 1980.
2. **Rosenkranz, H. S., McCoy, E. C., Sanders, D. R., Butler, M., Kiriazides, D. K., and Mermelstein, R.,** Nitropyrenes: isolation, identification, and reduction of mutagenic impurities in carbon black and toners, *Science*, 209, 1039, 1980.
3. **Richold, M., Jones, E., Rodford, R., and Clarke, D.,** Mutagenicity of typewriter ribbons: use of the Ames test as a biological screen, in *Mutagens in Our Environment*, Sorsa, M. and Vainio, H., Eds., Alan R. Liss, New York, 1982, 479.
4. Risk Assessment of Occupational Chemical Carcinogens, ECETOC Monogr. No. 3, Turner, L., Ed., European Chemical Industry Ecology & Toxicology Centre, Brussels, 1982.
5. **Astill, B. D., Lockhart, H. B., Moses, J. B., Nasr, A., Raleigh, R. L., and Terhaar, C. J.,** Tire testing in toxicology and decision-making criteria for their use, in *Mechanism of Toxicity and Hazard Evaluations*, Holmstedt, B., Lauwerys, R., Mercier, M., and Roberfroid, M., Eds., Elsevier, Amsterdam, 1980, 441.
6. **Hanley, J. W.,** Monsanto's "early warning" system, *Harvard Business Review*, November-December 1981, 107.
7. *Prudent Practices for Handling Hazardous Chemicals in Laboratories*, National Academy of Sciences Press, Washington, D.C., 1981.

Chapter 17

INDOOR AIR QUALITY AND EMISSIONS

R. A. Wadden

TABLE OF CONTENTS

I. THE OFFICE ENVIRONMENT

Recent concern about indoor air quality has been spurred by the increasing cost of energy. More extensive use of insulation, tighter building design to reduce infiltration, and increased ratios of recirculated/make-up air have all contributed to rising pollutant concentrations in interior spaces. The office environment has not ordinarily been considered a hazardous environment, but because of these energy considerations, and because types of office machines and procedures have changed significantly in the last 25 years (thereby changing the kinds of materials released into indoor air), exposure conditions can arise which constitute a hazard to public health.

Since perception of possible white-collar occupational problems is of recent note, relatively few hazards have been characterized. Of these, only a few have been identified as carcinogens or mutagens. In addition, the actual health effects due to relatively low exposure levels are often not well-defined. For instance the exposure of nonsmoking white-collar workers to cigarette smoke at work has been reported to cause decrements in the functioning of small airways in the lung.[1] What effect such decreases may have on overall health, or what effect the known carcinogens in cigarette smoke have on nonsmokers, are still matters of conjecture. Although the general health status of office workers is not well-defined it is likely that, excluding children, it would be similar to that of the population at large.[2] However, more persons who are susceptible to exposures to hazardous materials would be expected to work in offices than in occupations which traditionally involve polluted environments.

Potentially hazardous materials which have been identified include formaldehyde (particleboard and laminates, urea-formaldehyde insulation, carbonless paper), tobacco smoke, organics (spirit duplicators, carpet shampoos, copying machines), and ozone (copying machines, air cleaners). Asbestos and radon daughters from various types of construction materials have also been recognized as possible indoor hazards, although exposures are not specific to office environments. Some typical indoor concentrations are given in Table 1.

Not all of these substances are carcinogens. Tobacco smoke, at least for smokers, has been identified as a cancer hazard.[7] Levels of the carcinogenic benzo(*a*)pyrene and dimethylnitrosamine reported in Table 1 are essentially due to cigarette smoke. The studies suggesting that involuntary smoking may also contribute to cancer are summarized in Table 2, although consensus has yet to be reached on the cause/effect relationship.[3,8] Agencies responsible for worker health have recommended that formaldehyde be handled in the occupational setting as a potential carcinogen.[12] A variety of organic mutagens and suspect carcinogens have been measured in urban air,[13] but their possible enhancement in office environments has not yet been determined. Asbestos is recognized as an occupational cancer hazard, and asbestos fiber contamination of a building interior by fallout, contact or impact, and reentrainment has been reported.[14] Mutagenic impurities have been reported on some copying machine toners,[15] however, this has subsequently been traced to nitropyrene impurities in certain formulations of carbon black and in one case was eliminated by a process modification.[16,17] Many of the other pollutants in Table 1 are stressful to the lung and heart, and CO_2 is included because of its potential long-term influence toward reducing bone density.[18,19]

II. CHARACTERIZING THE HAZARD

A mass balance model on a pollutant released or penetrating into interior space is a convenient way to describe indoor concentrations. Conceptually,

$$\begin{matrix}\text{Pollutant}\\\text{flow}\\\text{in}\end{matrix} - \begin{matrix}\text{Pollutant}\\\text{flow}\\\text{out}\end{matrix} + \begin{matrix}\text{Source}\\\text{emissions}\end{matrix} - \begin{matrix}\text{Sink}\\\text{removals}\end{matrix} = \begin{matrix}\text{Indoor}\\\text{pollutant}\\\text{accumulation}\end{matrix} \qquad (1)$$

Table 1
TYPICAL POLLUTANT CONCENTRATIONS IN OFFICES, SCHOOLS, AND PUBLIC BUILDINGS[3]

Pollutants of concern	Concentration (sampling time)	Location
Carbon monoxide, CO	2.5-28 ppm	Offices, restaurants, bars, arenas
Respirable particles, RP		Restaurants, sports arenas, residences
	100-700 $\mu g/m^3$ (8-50 min)	With smoking
	20-60 $\mu g/m^3$ (1-42 min)	Without smoking
	10-70 $\mu g/m$ (24 hr)	Residences
Total suspended particulate matter	39-66 $\mu g/m^3$ (averages of 12-hr samples: 26-72% of outdoor concentrations)	Homes, public buildings
	2.7-79.4 $\mu g/m^3$ (48 hr)	Urban hospital
Asbestos	0-300 ng/m^3 (0-20 $\times$ 10^4 $fibers/m^3$) (5 min-10 hr)	Normal activities
	20 $\times$ 10^6 $fibers/m^3$	During maintenance
Formaldehyde, HCHO	60-1673 ppb (~1 hr; 463 ppb average for all measurements)	Homes with chipboard walls
	30-1770 ppb (35-60 min)	Mobile homes
	24-561 ppb; 41-3100 ppb (30-60 min)	Offices: schools (Holland)[4]
	196-448 (~2 hr)	Mobile day-care centers (Denmark)[5]
Ozone, O_3	<0.002-0.068 ppm (40 min to 2 hr)	Photocopying machine room
	<0.002-0.018 ppm (30 min)	Homes with electrostatic air cleaners
Benzo(*a*)pyrene	7.1-21.0 ng/m^3 (~2-4 hr)	Sports arena
Dimethylnitrosamine	0.11-0.24 $\mu g/m^3$ (90 min)	Bar
Methyl alcohol	365-3080 ppm (15 min, personal exposures)	Spirit duplicator areas in schools[6]
Carbon dioxide, CO_2	0.086% (5 min)	Lecture hall
	0.06-0.25%	School room
	0.9% (continuous measurements for ~8 wk)	Nuclear submarines
Viable particles	20-700 CFP/m^3 (av. of 10-min samples taken every 40 min)	Schools, hospitals, residences

One useful solution of Equation 1 is

$$C_i = \frac{k[q_o(1 - F_o) + q_2]C_o + S - R}{k(q_o + q_1F_1 + q_2)}[1 - e^{-(k/V)(q_o + q_1F_1 + q_2)t}] + C_s e^{-(k/V)(q_o + q_1F_1 + q_2)t} \quad (2)$$

where t is time; C is concentration indoors (C_i), outdoors (C_o), and at t = o (C_s); q is volumetric flow rate for make-up air (q_o), recirculation (q_1), and infiltration (q_2); F is filter efficiency for make-up (F_o) and recirculation air (F_1) (often the same); V is room volume; S is indoor source emission rate; R is indoor sink removal rate; and k, a factor which accounts

Table 2
HEALTH STUDIES OF INVOLUNTARY SMOKING AND CANCER[3]

Health study	No. of subjects (age)	Results
Nonsmoking Japanese wives	91,540 (≥ 40)	Wives of heavy smokers had significantly greater risk of developing lung cancer; age-occupation standardized annual mortality rates for lung cancer 8.7/100,000 for wives of occasional or nonsmokers; 14/100,000 for wives of ex-smokers or those smoking ≤ 19 cigarettes/day; 18.1/100,000 for wives of those smoking ≥ 20 cigarettes/day. The relative risk of passive smoking was about 1/3 to 1/2 that of direct smoking.[9]
Nonsmoking Greek women	189 (mean age = 62—63 years)	Statistically significant difference between cancer cases (40) and other patients (149) with respect to husband's smoking habits. Relative risk of lung cancer were 2.4 for those with husbands who smoked < 20 cigarettes/day and 3.4 for >20 cigarettes/day. Tentatively, the relative risk to passive smoking was 80% of that for direct smoking, but with broad confidence limits.[10]
Nonsmoking American women	176,739 (35—89)	Unadjusted lung cancer mortality ratios were 1.27 for women with husbands who smoked < 20 cigarettes/day and 1.10 for those married to ≥ 20 cigarettes/day smokers. When compared to those with nonsmoking husbands, neither mortality ratio was statistically significant. When data adjusted for age, race, educational status, residence, and husband's occupational exposure to dust, fumes, or vapors the mortality ratios were 1.37 for those with husbands smoking < 20 cigarettes/day and 1.04 for spouses of ≥ 20 cigarettes/day smokers, neither ratio being statistically significant.[11]

for inefficiency of mixing, is the fraction of incoming air which completely mixes within the room volume.[3]

Indoor concentrations may be estimated if the major components of Equation 2 are available. These include emission factors for S, deviations from perfect mixing ($k = 1$), infiltration and recirculation air rates, and outdoor concentrations. Often one or more of the terms may be neglected. Even if values of all the important terms cannot be determined it is often possible to use Equation 2 to estimate relative contributions to existing (measured) concentrations. Such estimates can serve as a reasonable basis for control strategy.

A variety of procedures are available to estimate infiltration.[3] Most of the methods require a measurement or estimate of the pressure difference across the building wall, ΔP. For a building envelope with no unusual openings, the pressure drop in Pa may be calculated from:

$$\Delta P = (0.6)(v^2) + (0.017)Pb\left[\frac{1}{T_o} - \frac{1}{T_i}\right] \qquad (3)$$

where v is the wind velocity in m/s, P the atmospheric pressure in Pa, b = building height in m, and T_o and T_i are outdoor and indoor temperatures, respectively, in degrees Kelvin. Use of Equation 3 in conjunction with infiltration tables[20] or Figure 1 supplies a value for infiltration.

Values of the mixing factor k have traditionally been chosen between 0.1 to 0.3. Slightly higher values (up to 0.6) may be justified for small rooms with considerable mixing from fans.

Pollutant removal by decay or deposition is often neglected for respirable particles (e.g., <2.5 μm in diameter) and nonreactive gases such as CO. Typical decay rates, K, for NO_2 (0.014/min), SO_2 (0.03/min), and O_3 (0.06/min) can be related to R with the following first-order form:

$$R = K\ V\ C_i \qquad (4)$$

More detailed information is given elsewhere.[3]

Emission factors (S) are summarized in Tables 3 (tobacco smoke), 4 (urea-formaldehyde

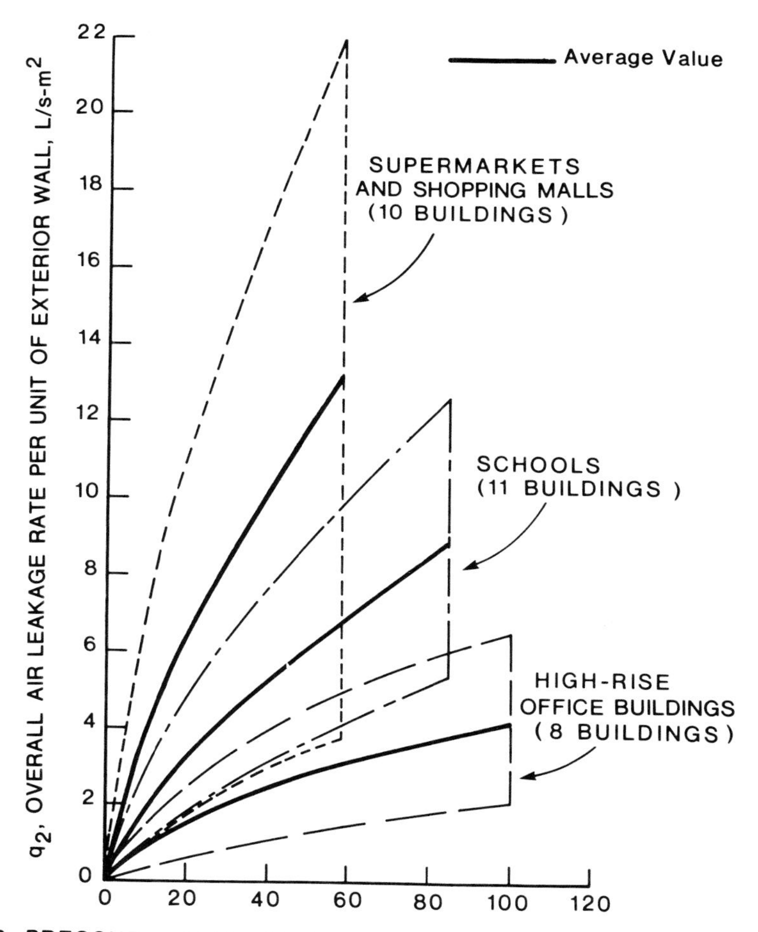

FIGURE 1. Comparison of overall air leakage for various building classifications. (From Shaw, C. Y., *ASHRAE J.*, 44, March 1981. With permission.)

insulation), 5 (office machines), and 6 (carbonless copy paper forms). In addition formaldehyde emissions from particleboard may be estimated from:

$$S' = \frac{(kn/\alpha)(RT - N)(aH + b)}{1 + nf/\alpha} \tag{5}$$

where S′ is in milligrams formaldehyde/(hr) (m^2 of board surface); k the fractional mixing efficiency; R = 0.064, a = 0.143, b = 0.048, f = 0.304, and N = 0.764 are empirical constants from the testing program; H is humidity, grams of H_2O/kg dry air; T is temperature, °C; n is air change rate per hour; α is area of board surface per volume of room, m^{-1}. The equation was found to be applicable for the following ranges: 17 to 32°C, 5-13 grams of H_2O/kg air, and 0.4 to 3 air changes per hour.[30] Other source and sink emissions factors may be found in Reference 3.

III. CONTROL OF INDOOR POLLUTANTS

Indoor pollution levels are ordinarily controlled by dilution with outdoor or cleaned

Table 3
EMISSION FACTORS FOR MAINSTREAM AND SIDESTREAM SMOKE[3]

Properties	Mainstream	Sidestream
General characteristics		
Duration of smoke production/cigarette	20 sec	550 sec
Amt. of tobacco burnt	347 mg	441 mg
No. of particles/cigarette	1.05×10^{12}	3.5×10^{12}
Particle no. median diameter	0.2 μm	0.15 μm
Particulate phase	**μg/cigarette**	**μg/cigarette**
Total suspended particulate matter	36,200	25,800
Tar (chloroform extract)	<500—29,000	44,100
Nicotine	100—2,500	2,700—6,750
Total phenols	228	603
Pyrene	50—200	180—420
Benzo(*a*)pyrene	20—40	68—136
Naphthalene	2.8	40
Methylnaphthalene	2.2	60
Aniline	0.36	10.8
Nitrosonornicotine	0.1—0.55	0.5—2.5
Cadmium	0.13	0.45
Nickel	0.08	—
Arsenic	0.012	—
2-Naphthylamine	0.002—0.028	0.08
Hydrogen cyanide	74	—
Polonium-210	0.029—0.044 pCi/cigarette	—
Gases and vapors	**μg/cigarette**	**μg/cigarette**
Carbon monoxide	1,000—20,000	25,000—50,000
Carbon dioxide	20,000—60,000	160,000—480,000
Acetaldehyde	18—1,400	40—3,100
Hydrogen cyanide	430	110
Methylchloride	650	1,300
Acetone	100—600	250—1,500
Ammonia	10—150	980—150,000
Pyridine	9—93	90—930
Acrolein	25—140	55—300
Nitric oxide	10—570	2,300
Nitrogen dioxide	0.5—30	625
Formaldehyde	20—90	1,300
Dimethylnitrosamine	10—65	520—3,380

recycled air, by collection of emissions, or by local exhaust of source discharges. Consensus ventilation standards have been developed by the American Society of Heating, Refrigerating, and Air Conditioning Engineers (ASHRAE) and many of these have been incorporated into national, state, and local building codes. One such standard, ASHRAE 90-75, has supplied the basis for many regulations dealing with building design to minimize energy use.[31] Another, ASHRAE 62-51, contains ventilation recommendations for minimizing indoor pollution. The suggested ventilation levels in this standard for various applications are based on restricting CO_2 concentrations to $\leq 2.5\%$, but do not specifically consider the levels or control of other pollutants. However, there is a recognition of the need for additional ventilation (up to 5 times the recommended rate) for areas with smoking. And there are

Table 4
FORMALDEHYDE EMISSIONS FROM UREA FORMALDEHYDE FOAMS

	Formaldehyde emission factor	
	µg/ (g foam · hr)	mg/ (m² foam surface · hr)
Average of data from 3 commercial foams tested at 33°C and 10 and 85% RH over interval of 10—30 days after foaming[22]	6.5	2.9
85% RH only	7.3	3.2
10% RH only	5.7	1.9
Commercial foam exposed at 35°C and 90% RH over interval 9—26 days[23]	10.2	—
Emissions determined at 25°C and 50% RH after 47-day purge following 16 mo. of storage[24]		
25°C		0.1—0.7
35°C (est.)		0.1—1.3

Table 5
OZONE EMISSIONS FROM OFFICE MACHINES AND ELECTROSTATIC AIR CLEANERS

	Maximum voltage	Ozone emission factors
		(µg/min)
2 Electric typewriters[25]	110	Not detectable; room conc. <0.002 ppm for 2 hr
		(µg/copy[a])
11 Photocopying machines[25,26]	3,500—11,000	Range <2—158; typically 15—45
		(µg/min)
Electrostatic air cleaners		
8 Installed in central air conditioning systems[25]	5,000—7,900	0—546
"Several well-known manufacturers' electronic air cleaners" (on central AC systems)[27]		303—1212
1 Portable unit[25]	9,900	84
2-Stage, low-voltage industrial unit (1 pass; 2 passes will double emission rate)[28]	11,000	333

[a] Typical copy rate, 5 copies/min.

some suggested indoor air quality goals as well. In general, the rate suggested for offices is from 2.5 to 5.0 ℓ/sec·person for nonsmoking areas.[32]

The use of recirculated air (which may constitute up to 80% of a building's recirculation load) requires careful consideration of the types of pollutants being generated and the methods proposed for their control. Filters effective for relatively large dust or lint particles will not be useful for cigarette smoke particles nor for the gaseous components of smoke. Activated charcoal is effective for the removal of carbon tetrachloride but not formaldehyde. In addition to appropriateness, reliability, maintenance, and mode of operation of control devices to clean recirculated air are necessary considerations when using this technique to minimize energy costs.

Removal of particulate matter is usually accomplished with filters or electrostatic precip-

Table 6
FORMALDEHYDE EMISSIONS FROM CARBONLESS COPY PAPER FORMS[29]

	Formaldehyde emissions		
No. of pages	μg/ form	μg/ kg form	μg/ $8^1/_2 \times 11$ in. sheet
3	0.41	39.1	0.11
4	0.45—1.0	66.2—68.8	—
5	0.95—13.2	53.5—714	0.18—2.60
6	17.7	858	4.20
8	4.6	154	0.68

itators. The cost and specific design of such systems depends on the particle size, required efficiency of removal, and the particle concentration. In general the smaller the average size particle to be collected, the greater the pressure drop through the appropriate filter, requiring greater blower capacity. Electrostatic precipitation does not cause as great a pressure drop, but the initial investment is usually greater, maintenance costs may be higher, and the device produces small but measurable quantities of ozone (Table 5). However, this method of control is quite effective for small particles.

Adsorption on activated carbon is the most common method of pollutant gas control. As with particles, the specific amounts of carbon required for control over a particular time period depend on the pollutant concentration and type, and the desired efficiency of control. A number of estimation techniques are reviewed in detail in Reference 3. Adsorbents are sometimes chemically coated to promote a reaction with pollutants which otherwise would not be removed. Such is the case with sodium sulfite-impregnated activated carbon which is effective for formaldehyde removal.

A major, but often overlooked, concomitant of pollution control systems is the necessity for regular and intelligent maintenance. Filters or adsorption beds are designed based on an assumed schedule of replacement and maintenance. All too often, such a schedule is not understood or followed, and the device is eventually ineffective and frequently by-passed.

REFERENCES

1. **White, J. R. and Froeb, H. F.,** Small airways dysfunction in non-smokers chronically exposed to tobacco smoke, *N. Engl. J. Med.*, 302, 720, 1980.
2. **National Center for Health Statistics,** Plan and Operation of the Second National Health and Nutrition Examination Survey, 1976-1980, *Public Health Service No. (PHS) 81-1317*, Department of Health, Education, and Welfare, Washington, D.C., July 1981.
3. **Wadden, R. A. and Scheff, P. A.,** *Indoor Air Pollution*, J. Wiley & Sons, New York, 1983.
4. **Van der Wal, J. F.,** Formaldehyde measurement in Dutch houses schools and offices in the years 1977-1980, *Atmos. Environ.*, 16, 2471, 1982.
5. **Olsen, J. H. and Dossing, M.,** Formaldehyde induced symptoms in day care centers, *Am. Ind. Hyg. Assoc. J.*, 43, 366, 1982.
6. **National Institute for Occupational Safety and Health,** Hazard Evaluation and Technical Assistance Report TA 80-32: Everett School District, Everett, Washington, NIOSH, Washington, D.C., June 1980.

7. **Surgeon General of the United States,** The Health Consequences of Smoking: Cancer, Department of Health and Human Services, Washington, D.C., 1982.
8. **Hammond, E. C. and Selikoff, I. J.,** Commentary: passive smoking and lung cancer with comments on two new papers, *Environ. Res.*, 24, 444, 1981.
9. **Hirayama, T.,** Non-smoking wives of heavy smokers have a higher risk of lung cancer: a study from Japan, *Br. Med. J.*, 282, 183, 1981.
10. **Trichopoulos, D., Kalandidi, A., Sparros, L., and MacMahon, B.,** Lung cancer and passive smoking, *Int. J. Cancer*, 27, 1, 1981.
11. **Garfinkel, L.,** Time trends in lung cancer mortality among non-smokers and a note on passive smoking, *J. Natl. Cancer Inst.*, 66, 1061, 1981.
12. **National Institute for Occupational Safety and Health,** Formaldehyde: Evidence of Carcinogencity, *Jt. Bull. NIOSH/OSHA*, No. 34, Washington, D.C., December 23, 1980.
13. **Singh, H. B., Salas, L. J., and Stiles, R. E.,** Distribution of selected gaseous organic mutagens and suspect carcinogens in ambient air, *Environ. Sci. Technol.*, 16, 872, 1982.
14. **Sawyer, R. N. and Spooner, C. M.,** *Sprayed Asbestos-Containing Material: A Guidance Document*, U.S. EPA Rep. No. EPA-450/2-78-014, Cincinnati, March 1978.
15. **Anon.,** Mutagens found in photocopying toners, *Chem. Eng. News*, 26, April 21, 1980.
16. **Rosenkranz, H. S., McCoy, E. C., Sanders, D. R., Butler, M., Kiriazides, D. K., and Mermelstein, R.,** Nitropyrenes: isolation, identification and reduction of mutagenic impurities in carbon black and toners, *Science*, 209, 1039, 1980.
17. **Löfroth, G., Hefner, E., Alfheim, I., and Moller, M.,** Mutagenic activity in photocopies, *Science*, 209, 1037, 1980.
18. **Schaeffer, K. E.,** Editorial summary (preventive aspects of submarine medicine), *Undersea Biomedical Res.*, 6(Suppl.), S-7, 1979.
19. **Tansey, W. A., Wilson, J. A., and Schaeffer, K. E.,** Analysis of health data from 10 years of Polaris submarine patrols, *Undersea Biomedical Res.*, 6(Suppl.), S-217, 1979.
20. *ASHRAE Handbook: 1981 Fundamentals*, American Society of Heating, Refrigerating and Air Conditioning Engineers, New York, 1981.
21. **Shaw, C. Y.,** Air tightness. Supermarkets and shopping malls, *ASHRAE J.*, 44, March 1981.
22. **Long, K. R., Pierson, D. A., Brennan, S. T., Frank, C. W., and Hahn, R. A.,** Problems associated with the use of urea-formaldehyde foam for residential insulation. I. The effects of temperature and humidity on formaldehyde release from urea-fomaldehyde foam insulation, University of Iowa, ORNL/SUB-7559/1, Oak Ridge National Laboratory, Department of Energy, Oak Ridge, Tenn., September 1979.
23. **Allan, G. G., Dutkiewicz, J., and Gilmartin, E. J.,** Long-term stability of urea-formaldehyde foam insulation, *Environ. Sci. Technol.*, 14, 1235, 1980.
24. **Hawthorne, A. R. and Gammage, R. B.,** Formaldehyde release from simulated wall panels insulated with urea-formaldehyde insulation, *J. Air Pollut. Control Assoc.*, 32, 1126, 1982.
25. **Allen, R. J., Wadden, R. A., and Ross, E. D.,** Characterization of potential indoor sources of ozone, *Am. Ind. Hyg. Assoc. J.*, 39, 466, 1978.
26. **Selway, M. D., Allen, R. J., and Wadden, R. A.,** Ozone emissions from photocopying machines, *Am. Ind. Hyg. Assoc. J.*, 41, 455, 1980.
27. **Sutton, D. J., Nodolf, K. M., and Makino, K. K.,** Predicting ozone concentrations in residential structures, *ASHRAE J.*, 21, September 1976.
28. **Holcomb, M. L. and Scholz, R. C.,** *Evaluation of Air Cleaning and Monitoring Equipment Used in Recirculation Systems*, NIOSH Publ. 81-113, National Institute for Occupational Safety and Health, Cincinnati, April 1981.
29. **Gockel, D. L., Horstman, S. W., and Scott, C. M.,** Formaldehyde emissions from carbonless copy paper forms, *Am. Ind. Hyg. Assoc. J.*, 42, 474, 1981.
30. **Anderson, I., Lundquist, G. R., and Molhave, L.,** Indoor air pollution due to chipboard used as a construction material, *Atmos. Environ.*, 9, 1121, 1975.
31. *ASHRAE Standard: Energy Conservation in New Building Design*, ASHRAE 90-75, American Society of Heating, Refrigerating and Air Conditioning Engineers, New York, 1975.
32. *Standards for Ventilation Required for Minimum Acceptable Indoor Air Quality*, ASHRAE 62-81, American Society of Heating, Refrigerating and Air Conditioning Engineers, New York, 1981.

Chapter 18

EVALUATION OF OFFICE MATERIALS FOR GENOTOXIC EFFECTS

June J. Andersen, Ann D. Burrell, Gary M. Decad, and Betty J. Dabney

TABLE OF CONTENTS

I. STRATEGY FOR STUDIES IN VITRO AND IN ANIMALS

A. The Office Environment

Because the sciences of toxicology and analytical chemistry have advanced so rapidly in the past decade, the toxicologist now has a powerful battery of techniques to use in the investigation of materials to ensure their safe use in the office. An awareness of the potential for exposure to genotoxic material in the office environment has developed chiefly during the past 5 years. Short-term screening tests like the Salmonella/microsome assay[1] have contributed to that awareness since we are now able to screen large numbers of materials. Another technology change is the advancement of measurement techniques such as sorption chromatography and mass spectroscopy which allow detection of genotoxic materials in concentrations less than the microgram per cubic meter range.

In devising a strategy to assure office safety we must consider materials from three categories. The first category is the environment itself, which would include the air conditioning system, the materials used in furniture, walls and carpets, cigarettes, and the lighting system. The second category is the equipment now in common use in a high technology office environment which includes electronic typewriters, video display terminals, personal computers, printers, and electrophotographic copiers. The third category is the more conventional materials and supplies such as paper, pens, typewriter ribbons, ink pads, correction fluids, and typewriter cleaners.

The evaluation of this equipment and material for toxic effects often begins with acute toxicity and genotoxicity screening assays since the chemicals used in a high technology environment may be unique and no data exists in the literature as a starting point. Even chemicals such as dyes and solvents may not be referenced in the literature, since they were never considered for use in food, drugs, or pesticides or may have been considered to be proprietary by the manufacturer.

The office environment itself offers some unique features compared to other occupational exposures. The worker population is very large, and is exposed to products from a wide and variable number of manufacturers. The exposure levels are normally very low and difficult to measure, although measurable exposures may occur in cases of materials misuse. Since the exposure levels are very low, the office environment is generally considered to be safe relative to the risks which may be encountered in other occupations.

B. Strategy of Toxicology Studies for Office Materials

We have developed a strategy for evaluation of chemicals used in office products (Figure 1). Concurrent with generating a material toxicity profile, the product quantity and life are determined, followed by materials balance estimates during use in the product environment. If required, quantitative measurements are made for emissions due to wear, outgassing, or thermal emissions. The size and composition of the potentially exposed population is also determined. This population may include both users of the equipment and service personnel who repair the equipment. The exposure levels are often very different for these two classes of individuals. Finally, the relevant route of human exposure must be defined and is typically the inhalation route for exposure to airborne contaminants, or the dermal route for exposures to inks, papers, cleaners, etc.

The toxicity profile is generated by a review of the literature, followed by testing as necessary to establish the acute toxicity, skin and eye irritation effects, skin sensitization, and the potential for genotoxicity and reproductive effects. Short-term assays for genotoxicity in both bacterial and mammalian cell systems are used to detect the potential for chronic effects such as mutation or cancer. The number and types of assays used are guided both by the chemical structure and the intended use. The tier structure of the tests utilized is consistent in concept with the safety decision tree developed by the Food Safety Council,[2]

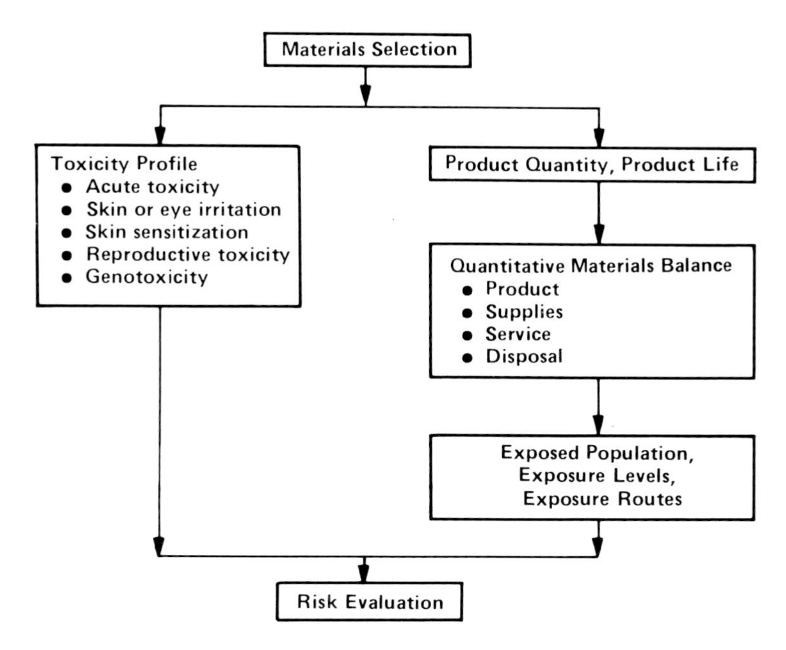

FIGURE 1. Strategy for evaluation of office materials for genotoxic effects.

with additional emphasis on the physical and chemical properties of the material being considered and the relevant route of exposure.

Finally the data are reviewed and a decision is made as to whether the data are adequate for a risk evaluation given the route and level of potential human exposure. If the data are inadequate, a test plan must be formulated to pursue a second tier of tests including evaluation of whether the material is absorbed via a relevant route, how the material may be metabolized, and even chronic animal studies if required. Before such expensive and time-consuming tests are pursued, however, the substitution of a less toxic material will be explored, since this alternative is always preferable. If the data are adequate, a qualitative or quantitative risk evaluation is made to ensure that the material can be safely used in the office.

II. SHORT-TERM TEST UTILITY IN THE EVALUATION OF TYPEWRITER RIBBONS

A. Identification of Mutagenic Contaminants

One example of short-term test utility is the ability to identify the source of mutagenic materials in a finished product.

In 1981, Dr. Göran Löfroth reported that extracts of typewriter ribbons can be mutagenic in the Salmonella/microsome assay.[3] This assay, also known as the Ames test measures the ability of a chemical or mixture to cause mutations in special tester strains of *Salmonella typhimurium*.[1] It is the most widely used short-term in vitro test in toxicology today, and there is a 70 to 80% correlation between the activity of a chemical in the Salmonella/microsome assay and its carcinogenicity in laboratory animals.

Dr. Löfroth's results were confirmed with a more detailed study of IBM ribbons. Mutagenic studies of typewriter ribbon extracts are complicated by complex formulations of ribbon inks, and the ingredients themselves may be mixtures of unknown composition and purity.[4,5] This is especially true of the many dyes and pigments used in inks. In addition, there may be considerable differences in the same kinds of ribbons manufactured in different locations. Thus, there may be substantial differences between ribbon lots.

Table 1
MUTAGENIC ACTIVITY OF HIGH-YIELD CORRECTABLE RIBBON EXTRACTS IN THE SALMONELLA/MICROSOME ASSAY (FOLD INCREASE OVER BACKGROUND, AVERAGE OF 3 LOTS IN STRAIN TA98)

Ribbon source	No metabolic activation	S9 Metabolic activation
Old European	1.3 (S.D. 0.5)	1.1 (S.D. 0.3)
New European	1.3 (S.D. 0.8)	1.1 (S.D. 0.1)
Old American	5.0 (S.D. 4.6)	1.5 (S.D. 0.3)
New American	1.2 (S.D. 0.3)	1.1 (S.D. 0.2)

Carbon black is a common ribbon ingredient. Nitrated carbon blacks manufactured prior to 1981 may have contained several hundred parts per million of nitropyrenes, some of the strongest bacterial mutagens known.[6] In 1981, the levels of nitropyrene impurities in carbon blacks were drastically reduced. Because Dr. Löfroth's studies were done before this change in nitropyrenes, it is likely that his results may have been influenced by the nitropyrenes in the carbon blacks.

Mutagenic activity was assayed in extracts of black fabric, high-yield correctable and noncorrectable film ribbons. We compared the mutagenic activity in equivalent ribbons manufactured in the U.S. and abroad, both before and after the change in the nitropyrene levels in three different lots of each kind of ribbon. We initially used Löfroth's method, in which 20 cm of the ribbon is extracted in 5 mℓ of dimethylsulfoxide (DMSO). This is a rather harsh treatment, however, and is designed to determine if there is any mutagenic activity present in the sample. We were also interested in knowing how accessible any mutagenic activity might be in more relevant solvents. Finally, our plan was to identify the sources of any mutagenic activity in the ribbons, and to reformulate the ribbons where possible.

All of the results are expressed as fold-elevation over background in strain TA98 only, which was determined to be the best responding bacterial strain. Generally, the results are regarded as positive if there is at least a two fold elevation over background. No appreciable activity was seen in DMSO extracts from the high-yield correctable ribbons, either for the European or American ribbons, manufactured after the changeover in carbon blacks (Table 1.) While there was some activity seen in the old American ribbons, the large standard deviation seen for this type of ribbon indicates much lot-to-lot variability. Notably, none of the current high-yield correctable ribbons was mutagenic when tested under these conditions.

For the fabric ribbons, both the European and American versions contained mutagenic activity if they were manufactured prior to 1981, but the current ribbons are not mutagenic (Table 2). This pattern of the loss of mutagenic activity, coinciding with the reduction of nitropyrenes in the carbon blacks, would suggest that the activity seen in the older ribbons was due to the nitropyrenes. Also, the profile of activity in these older ribbons, with no exogenous metabolic activation required, is consistent with behavior for nitropyrenes.

The noncorrectable film ribbon, however, showed a different result (Table 3). Two of the three European ribbon lots were highly mutagenic regardless of the date of manufacture, while the American ribbons were not mutagenic. We have confirmed that of all the ingredients in the European noncorrectable film ribbon, only the carbon black is mutagenic. Furthermore, the "old" European ribbon was made with a nitrated carbon black, while the carbon black used to make the American ribbon was not nitrated.

Table 2
MUTAGENIC ACTIVITY OF FABRIC RIBBON EXTRACTS IN THE SALMONELLA/MICROSOME ASSAY (FOLD INCREASE OVER BACKGROUND, AVERAGE OF 3 LOTS IN STRAIN TA98)

Ribbon source	No metabolic activation	S9 Metabolic activation
Old European[a]	3.2 (S.D. 0.6)	3.2 (S.D. 1.2)
New European	1.0 (S.D. 0.4)	1.5 (S.D. 0.5)
Old American	28.4 (S.D. 2.6)	13.1 (S.D. 2.9)
New American	1.2 (S.D. 0.3)	1.8 (S.D. 0.5)

[a] Two lots tested.

Table 3
MUTAGENIC ACTIVITY OF NONCORRECTABLE FILM RIBBONS IN THE SALMONELLA/MICROSOME ASSAY (FOLD INCREASE OVER BACKGROUND, IN STRAIN TA98)

Ribbon source	No. of lots	No metabolic activation	S9 Metabolic activation
Old European	3	74.0 (S.D. 26.2)	15.4 (S.D. 10.9)
New European	2	67.5 (S.D. 9.7)	14.0 (S.D. 2.9)
	1	1.9 (S.D. 1.0)	1.8 (S.D. 0.1)
Old American	3	0.8 (S.D. 0.2)	1.5 (S.D. 0.4)
New American	3	0.9 (S.D. 0.4)	1.2 (S.D. 0.2)

However, if the nitropyrenes in the carbon black were the source of the mutagenic activity, why weren't the "new" European ribbons less active? We found that the European carbon blacks were purchased through a broker, thereby incurring considerable delay between the date of manufacture of the carbon black and its date of use. More recent European noncorrectable film ribbons, made with carbon blacks known to contain much lower levels of nitropyrenes, have not been mutagenic when tested by this method.

B. Accessibility of Mutagenic Activity

To determine how accessible any mutagenic activity might be under more relevant conditions, we have examined extracts of the mutagenic noncorrectable film ribbon in a variety of solvents. These results are shown in Table 4. Much less mutagenic activity was extracted in physiological saline, in hand lotion, or in a synthetic sebum (skin oil) mixture, than in DMSO. In the last condition shown, an extract in skin oil was subsequently re-extracted in DMSO, to demonstrate that the artificial sebum had not interfered with the assay.

Over the past few years, we have systematically tested the entire IBM ribbon inventory for mutagenic activity. The new IBM ribbons are also being scrutinized very thoroughly as they are developed, both for mutagenic activity, and for possible dermal irritation and sensitization as well. We are also conducting a similar testing program on inks and papers.

At this point, we should emphasize that while the Salmonella assay is of great utility as an initial screening tool for genotoxic activity, the significance of this activity for human

Table 4
MUTAGENIC ACTIVITY OF NONCORRECTABLE FILM RIBBON EXTRACTS (NO METABOLIC ACTIVATION)[a]

Solvent	TA98
DMSO	69.9
Saline	3.4
Hand lotion	7.2
Skin oil	15.8
Skin oil/DMSO	13.9

[a] Fold increase over background.

FIGURE 2. Disposable ribbon cartridge.

exposure is unknown. The actual hazard to humans from any ribbon must be insignificant, since there is no substantial exposure to ribbons either by direct contact or from ink on the printed page.

Human exposure also depends on the ribbon packaging. For example, snap-in machine configurations or disposable ribbon cartridges (Figure 2) eliminate skin contact. Nevertheless, we assume that any ribbon made from nonmutagenic components is preferable and ribbon reformulation would be explored.

III. STRATEGY FOR EVALUATION OF A WIDELY USED PHOTOCONDUCTOR

A. Photoconductor Structure and Function

The electrophotographic process is used for copying and printing in several IBM products. The photoconductor is the key element in this process, and the structure of a layered photoconductor is shown in Figure 3. Light is used to generate a charged pattern on the surface of the photoconductor which attracts toner. The pattern of toner is transferred to paper, fused, and the result is a printed page. The materials in the top layer of the photoconductor are of interest because very small amounts of this layer can be removed from the

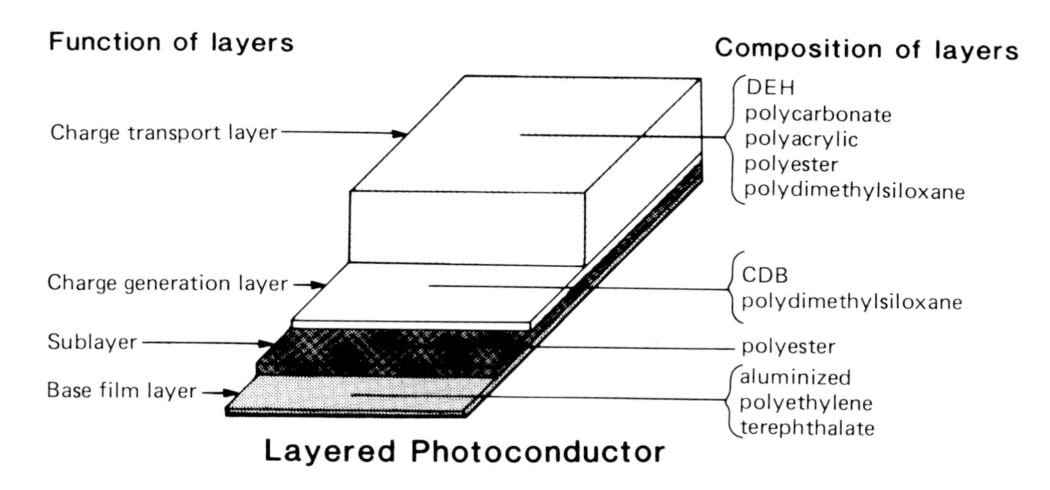

FIGURE 3. Structure of a layered photoconductor.

	Oral LD_{50}	Dermal LD_{50}	Inhalation LC_{50}	Skin irritation	Eye irritation
Intact photoconductor	Not tested	Not tested	Not tested	Non-irritant	Not tested
Photoconductor dust	Not tested	Not tested	Nontoxic 16.2 mg/m^3/4hr	Non-irritant	Non-irritant
DEH	>10 g/kg	>2.0 g/kg	Nontoxic 49.4 mg/m^3	Non-irritant	Non-irritant
CDB	>3.16 g/kg	>3.16 g/kg	Not tested	Non-irritant	Slight irritant 39 mg/eye
Polyester	Approved by FDA for use in medical devices and food containers.*				
Polycarbonate	Approved for sale as a food-grade packaging material by FDA.*				
Polyacrylic	High-molecular-weight polymer.				

*Manufacturers' data and correspondence.

FIGURE 4. Acute toxicology of photoconductor and components.

photoconductor by abrasive wear and deposited in spent toner, transferred to the printed page, or discharged into the air.

B. Basic Toxicity Profile

The initial toxicity profile of both the photoconductor and its components was conducted using the Federal Hazardous Substances Act Guidelines. The results in Figure 4 showed that the photoconductor and its components were neither toxic nor irritating using relevant routes of administration at levels many times the potential exposure concentration. Initial Salmonella/microsome assays of new and used photoconductor itself and of the major components showed no evidence of mutagenic activity.[7]

p-Diethylaminobenzaldehyde-diphenylhydrazone (DEH)

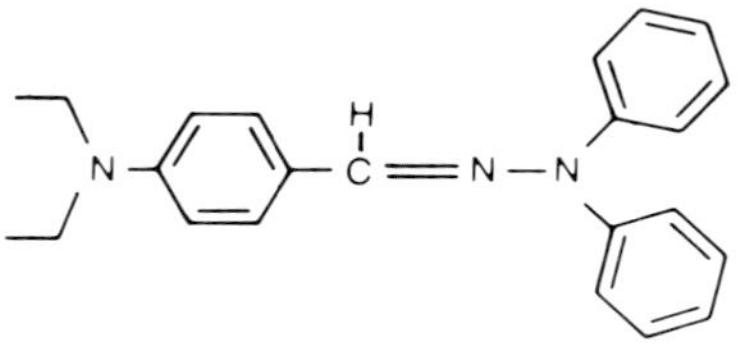

Chlorodiane Blue (CDB)

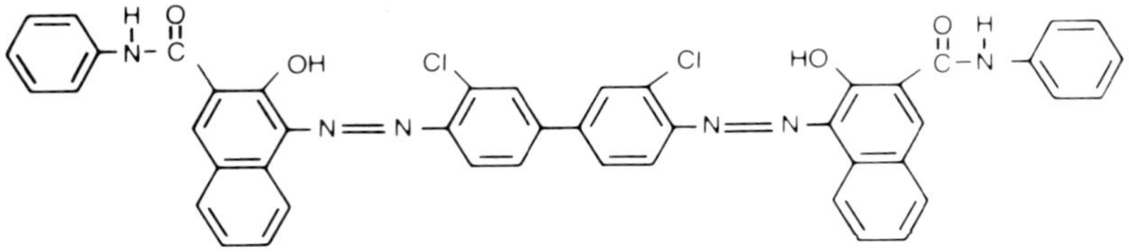

FIGURE 5. Structure of DEH and CDB.

C. In Vitro and In Vivo Assays of Genotoxic Potential

Two components were selected for further study because their usage was unique to this application and no literature data was found. *p*-Diethylaminobenzaldehyde diphenylhydrazone (DEH) (Figure 5) performs the charge transport function in the top layer of the photoconductor. DEH was assayed using two mammalian cell assays in vitro for genotoxicity. The forward mutation assay at the X-linked *hgprt* (hypoxanthine guanine phosphoribosyltransferase) locus[8] in Chinese hamster ovary (CHO) cells showed no indication of a response at any dose tested up to solubility limits of 400 μg/mℓ. In addition, unscheduled DNA synthesis was assayed in primary rat hepatocytes as another indicator of potential genotoxicity.[9] The use of primary hepatocytes is particularly useful where metabolic products generated by the liver might be genotoxic. Concentrations of DEH up to 0.1 mg/mℓ gave no indication that DNA repair had been stimulated.[7] Further testing of DEH was not necessary since operators of copiers do not come in contact with photoconductors and because the surface is resistant to abrasive wear.

The charge generation pigment Chlorodiane Blue (CDB) (Figure 4) was also selected for testing because it is derived from the known carcinogen 3,3′-dichlorobenzidine.[10] CDB itself was nonmutagenic in Salmonella, but as expected, its reduction products were mutagenic after chemical reduction in vitro with dithionite. Reduction probably liberated dichlorobenzidine, which is known to be mutagenic in Salmonella (Figure 6).

CDB was also assayed for its ability to stimulate unscheduled DNA synthesis in primary rat hepatocytes. Riboflavin has been shown to stimulate reductive cleavage of some azo dyes in vitro. Even in the presence of riboflavin, the pigment did not stimulate DNA repair at concentrations from 1 μg to 1 mg/mℓ.[7]

Some carcinogenic azo dyes give false negative results in vitro assays for genotoxicity. In animals they can be reductively cleaved by gut bacteria. In order to rule out this possibility, CDB was radiolabeled using [^{14}C]3,3′-dichlorobenzidine as a synthetic precursor. The [^{14}C]-labeled pigment Pigment Yellow 12 (PY12) was also prepared. PY12 is a dichlorobenzidine-derived pigment which showed no evidence of carcinogenicity after lifetime ingestion in rodents and no absorption or metabolism after oral administration to rabbits. Since PY12 is a chemical congener of CDB with similar physical properties, we expected them to be similar in the degree of absorption and metabolism.

The pigments were administered by oral and dermal routes of exposure to Fischer 344

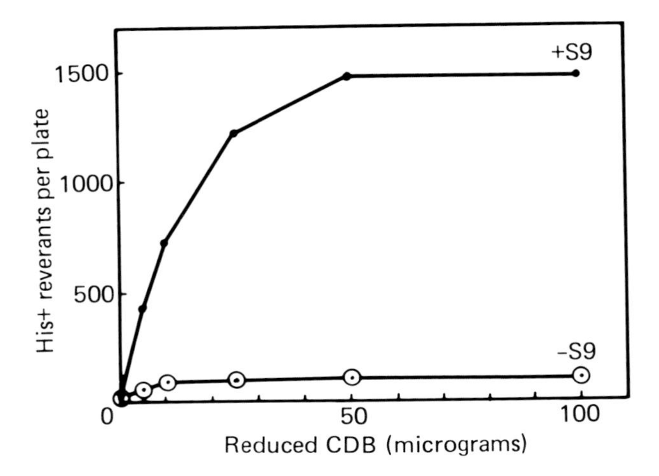

FIGURE 6. Mutagenicity of CDB in *Salmonella typhimurium* after chemical reduction with dithionite.

rats. Radioactive doses were sufficient to detect 0.001% of the dose in tissue, blood, or excreta. The pigments were not absorbed by either route. Furthermore, after oral dosing all of the dose was recovered in the feces and cochromatographed with the unlabeled pigments.[11] Both of these pigments are extremely water insoluble which may account for their resistance to reductive cleavage in the gut.

Further testing of CDB was not necessary since we could conclude that it showed a low potential for genotoxicity in vivo. In addition, it is buried beneath the charge transport layer of the photoconductor (Figure 3) and would not be liberated by abrasive wear.

IV. STRATEGY FOR MATERIALS WITH SIGNIFICANT HUMAN EXPOSURE

The strategy for evaluating the genotoxic properties of a material with significant human exposure would differ from the ones described here.

Positive results in in vitro screening assays would indicate the need for subchronic and chronic animal bioassays to measure the carcinogenic or mutagenic potency. In fact, such animal studies are rarely pursued because of the high cost and lengthy time required for completion. Rather, alternative materials would be explored which have a lower potential for genotoxic activity based on short-term screening assays or dermal absorption studies.

Quantitative risk assessment is a useful tool for estimating the level of concern when animal studies do exist which give dose response information on the carcinogen potency of a chemical.[12] Although the currently available methods have debatable significance, calculations can be made using conservative models which bound the upper value of the possible risk.[13] The exposure levels for materials used in the office environment are frequently so low that the calculated upper bound risk is insignificant. Responsible public education is needed on carcinogenic risks from chemicals and materials so that office workers are not unduly alarmed by insignificant risks and can place these risks in proper perspective relative to those experienced in everyday life.

For materials already in use in the office, retrospective epidemiological studies are appropriate when new information is discovered about a material which has had a long history of use. However, it is difficult to identify a population with sufficient exposure time to one specific material and such studies can be easily confounded by personal habits such as smoking.

Finally, the evaluation of office materials for genotoxic potential must be ongoing, since the test technologies and measurement capabilities are rapidly evolving. Strategies for studies in vitro and in animals must also be constantly evolving to reflect the state of the art in testing and to assure the safe use of materials in the office.

ACKNOWLEDGMENTS

Some of the experiments reported here were performed in the laboratories of K. Mortelmans, A. Mitchell, and C. Mitoma, SRI International; J. Carver, Battelle Columbus; and G. Williams, American Health Foundation. In addition to the authors, experimental results from F. Joachim, C. Snyder, F. Aldrich, and W. Bocim, IBM, are acknowledged, with special appreciation to J. Harris for his leadership and support.

REFERENCES

1. **Ames, B. N., McCann, J., and Yamasaki, E.,** Method for detecting carcinogens and mutagens with the *Salmonella*/mammalian microsome mutagenicity test, *Mutat. Res.*, 31, 347, 1975.
2. Food Safety Council, A proposed food safety evaluation process. Final report of the board of trustees, The Nutrition Foundation, New York, June 1982.
3. **Moller, M., Alfheim, I., Löfroth, G., and Agurell, E.,** Mutagenicity of extracts from typewriter ribbons and related items, *Mutat. Res.*, 119, 239, 1983.
4. **Miloy, P. and Kay, K.,** Mutagenicity of nineteen major graphic arts and printing dyes, *J. Toxicol. Environ. Health*, 4, 31, 1978.
5. **Kay, K.,** Toxicological and cancerogenic evaluation of chemicals used in the graphics arts, industries, *Clin. Toxicol.*, 9, 359, 1976.
6. **Mermelstein, R. et al.,** The extraordinary mutagenicity of nitropyrenes in bacteria, *Mutat. Res.*, 89, 187, 1981.
7. **Burrell, A. D. and Andersen, J. J.,** *In vitro* toxicity testing of chlorodiane blue-based photoconductor and its components (Abstract), Environmental Evaluation of Office Materials for Genotoxic Effects Mutagen Society, Bethesda, Md., 1982.
8. **Carver, J. H., Adair, G. M., and Wandres, D. L.,** Mutagenicity testing in mammalian cells. II. Validation of multiple drug resistance markers having practical application for screening potential mutagens, *Mutat. Res.*, 72, 207, 1980.
9. **Williams, G. M.,** Detection of chemical carcinogens by unscheduled DNA synthesis in rat primary cell cultures, *Cancer Res.*, 37, 1845, 1977.
10. International Agency for Research on Cancer, *Monographs on the Evaluation of Carcinogen Risk to Man: 3,3'-dichlorobenzidine*, Vol. 4, IARC, Lyon, 1974, 49.
11. **Decad, G. M., Snyder, C. D., and Mitoma, C.,** Fate of water-soluble and water-insoluble dichlorobenzidine-based pigments in Fischer 344 rats, *J. Toxicol. Environ. Health*, in press.
12. **Van Ryzin, J.,** Quantitative risk assessment, *J. Occup. Med.*, 22, 321, 1980.
13. Office of Technology Assessment, Assessment of technologies for determining cancer risks from the environment, PB81-235400, June 1981.

Index

INDEX

A

B

C

D

E

F

G

H

I

J

K

L

M

N

O

P

Q

R

S

T

U